Springer Theses

Recognizing Outstanding Ph.D. Research

For further volumes:
http://www.springer.com/series/8790

Aims and Scope

The series "Springer Theses" brings together a selection of the very best Ph.D. theses from around the world and across the physical sciences. Nominated and endorsed by two recognized specialists, each published volume has been selected for its scientific excellence and the high impact of its contents for the pertinent field of research. For greater accessibility to non-specialists, the published versions include an extended introduction, as well as a foreword by the student's supervisor explaining the special relevance of the work for the field. As a whole, the series will provide a valuable resource both for newcomers to the research fields described, and for other scientists seeking detailed background information on special questions. Finally, it provides an accredited documentation of the valuable contributions made by today's younger generation of scientists.

Theses are accepted into the series by invited nomination only and must fulfill all of the following criteria

- They must be written in good English.
- The topic should fall within the confines of Chemistry, Physics, Earth Sciences, Engineering and related interdisciplinary fields such as Materials, Nanoscience, Chemical Engineering, Complex Systems and Biophysics.
- The work reported in the thesis must represent a significant scientific advance.
- If the thesis includes previously published material, permission to reproduce this must be gained from the respective copyright holder.
- They must have been examined and passed during the 12 months prior to nomination.
- Each thesis should include a foreword by the supervisor outlining the significance of its content.
- The theses should have a clearly defined structure including an introduction accessible to scientists not expert in that particular field.

Dhriti Sundar Ghosh

Ultrathin Metal Transparent Electrodes for the Optoelectronics Industry

Doctoral Thesis accepted by
ICFO-The Institute of Photonic Sciences,
Barcelona, Spain

Springer

Author
Dr. Dhriti Sundar Ghosh
Optoelectronics
ICFO-The Institute of Photonic Sciences
Mediterranean Technology Park
Castelldefels, Barcelona
Spain

Supervisor
Prof. Dr. Valerio Pruneri
Optoelectronics
ICFO-The Institute of Photonic Sciences
Mediterranean Technology Park
Castelldefels, Barcelona
Spain

ISSN 2190-5053 ISSN 2190-5061 (electronic)
ISBN 978-3-319-34414-0 ISBN 978-3-319-00348-1 (eBook)
DOI 10.1007/978-3-319-00348-1
Springer Cham Heidelberg New York Dordrecht London

Printed on acid-free paper

Springer is part of Springer Science+Business Media (www.springer.com)

Dedicated to my mom

Supervisor's Foreword

Transparent electrodes (TEs) are essential elements in many photonic and electronic devices since they allow applying or collecting electrical signals while transmitting electromagnetic radiation.[1] Devices such as tactile and rigid displays, light emitting diodes, solar cells, electrochromic windows, and photodiodes would not exist if suitable TEs were not available.

Currently, the most widely used TE is indium tin oxide (ITO), a wide band gap semiconductive oxide, which presents the best trade-off between optical transparency and electrical conductivity and that was also developed many decades ago.[2] ITO suffers from important drawbacks; it is expensive because Indium is a scarce element, its price is subjected to significant variations leading to unwanted change in production costs, it is not always chemically compatible with other layers forming the devices, such is the case of organic compounds that can be affected by the release of In ions, and it has poor mechanical flexibility because of its fragile oxide nature, this property being essential for low cost roll-to-roll processing.[3]

There is thus a strong need to develop cost-effective, mechanically flexible and, in some cases, chemically stable TEs which can replace ITO. Several research groups have been investigating a wide range of Indium-free materials, including aluminum-doped zinc oxide (AZO), carbon nanotubes (CNTs), silver nanowires (AgNWs), ultrathin metal films (UTMFs) and, and most recently, graphene.

Dhriti Sundar Ghosh's thesis contains original work on UTMF-based TEs, covering their design, fabrication, characterization, and effective use in several devices, in particular organic solar cells (OSCs) and organic light-emitting diodes (OLEDs). Metals, when sufficiently thin, i.e., <10 nm, can become transparent while still maintaining high electrical conductivity. This is the case for example of Ni, Ti, Cr, Cu, and Ag. The development of UTMF technology within this thesis has allowed to produce high quality In-free TEs, either in single or multilayer

[1] Handbook of transparent conductors, ISBN 978-1-4419-1638-9, Springer (2010).

[2] Method of fabricating transparent conductors, US3749658 (1973).

[3] Past achievements and future challenges in the development of optically transparent electrodes, Nature Photonics 6, 809 (2012).

geometry, as well as their combination with other materials, such as AZO, CNTs, and, more recently, graphene.

Main achievements of the thesis include: for the first time it is shown that through ad hoc surface treatment, the island growth of metal during deposition can be avoided, thus making possible to achieve UTMFs which are continuous even at 1 nm thickness, i.e., only several atomic layers[4]; oxidation which is usually an unwanted effect is exploited to increase the stability of the UTMFs together with their transparency and, in some cases, the work function for carrier injection[5]; UTMFs are combined with metallic grid for achieving a better trade-off between transparency and conductivity, the methods being readily extendable to Ag NWs and CNTs networks[6]; UTMFs are used to increase the stability of AZO, especially under humidity and temperature conditions, while, at the same time, providing them with functional surface properties.[7] Last, but not least, the thesis demonstrates that the developed UTMF-based TEs can be efficiently used in OSCs and OLEDs, providing performance matching those of similar ITO-based devices, with the additional advantages of low cost and mechanically flexibility.[8,9]

The scientific and technological advances, covering material, surface chemistry, and device physics, are described in 12 papers in high impact factor journals (5 of them as first author), 5 invited talks, and 3 patent applications. Arguably, the most important output of the thesis's work is the fact that it has allowed the group to establish numerous bilateral projects within industries in Spain, Germany, Italy, Israel, and USA with prominent companies like Ficosa International, SEAT-Volkswagen, BASF, Carl Zeiss, Oclaro, IBM, and Corning Inc., for a wide range of applications, including tactile and rigid displays, electro-optic modulators, OSCs, lithography masks, and car equipment.

Barcelona, March 2013 Valerio Pruneri

[4] Widely transparent electrodes based on ultrathin metals, Optics Letters 34, 325 (2009).

[5] Stable transparent Ni electrodes, Optical Materials 31, 1115 (2009).

[6] High figure-of-merit ultrathin metal transparent electrodes incorporating a conductive grid, Applied Physics Letters 96, 41109 (2010).

[7] Highly stable Al-doped ZnO transparent conductors using an oxidized nickel capping layer, Applied Physics Letters 99, 93302 (2011).

[8] Semitransparent metal electrode of Cu-Ni as a replacement of ITO in organic photovoltaic cells, Solar energy Materials, and Solar Cells 95, 1228 (2011).

[9] Nickel as an alternative semitransparent anode to indium tin oxide for polymer LED applications, Nanotechnology 25, 275204 (2009).

Preface

Transparent electrodes (TEs) are the essential elements of many optoelectronic devices such as solar cells, touch screens, organic LEDs, and LCDs. Consequently, demand for TEs is growing very steeply and the market value presently stands at eight billion USDs. The state-of-art indium tin oxide (ITO) has an excellent trade-off between optical transparency and electrical sheet resistance but suffers from several drawbacks, mainly the increasing cost due to indium shortage, and inadequate flexibility due to poor mechanical ductility.

This thesis presents the development of a new class of TEs based on ultrathin metal films (UTMFs). The work started from understanding the fundamental aspects of UTMF growth and properties, and then focused on different UTMF-based geometrics, composition, and combination for potential applications in different optoelectronic applications.

Single component ultrathin Ni and Cr films were shown to possess significantly high transparency in the ultraviolet (175–400 nm) and mid-infrared (2.5–25 μm) regions making them viable TE for devices such as UV photodiodes, and IR pyroelectric detectors. The natural oxidation process, which is a major concern for metal films, has been exploited to achieve stable metallic films by inducing a protective oxide layer.

In another proposed novel design, incorporating an ad hoc conductive grid, the sheet resistance of UTMFs can be reduced by more than two orders of magnitude with negligible loss in transparency, which in turn eliminates the inverse trade-off relationship between optical transparency and electrical conductivity of continuous metal-based TEs.

A TE structure based on the ultrathin conductive Cu films with an application specific functionalized capping layer of Ti or Ni layer has been demonstrated. The properties of the TE can be tuned accordingly and show excellent stability against temperature, and oxidation. The suitability of Ag–Cu alloy films as TE as an alternative to ITO has been also investigated. The optical spectrum of such alloy films follows the average optical behavior of single component Cu and Ag layers, thus resulting in a much flatter optical response in the visible region.

UTMFs combined with Al-doped ZnO (AZO), which is possible ITO replacement, has also been demonstrated to show the possibility of hybridizing the two technologies. A bilayer Ag/AZO has been developed which can overcome the high reflection of metals and retain their good electrical behavior, while maintaining a minimum total film thickness. In another structure, UTMF capping layer were used to improve the stability of AZO. It was found that an ultrathin oxidized Ni capping layer with a thickness at percolation threshold greatly enhances the stability of AZO layer in harsh environment without affecting the electro-optical properties.

Contents

Chapter 1
Introduction

1.1 Transparent Electrodes

One of the most crucial requirements of optoelectronics and photonics industry is the realization of inexpensive and efficient transparent electrodes (TEs), i.e., films that permit one to bring electrical currents or potentials in the proximity of optically active regions without significant loss of optical energy [1, 2]. For example, in case of photovoltaic (PV) cells, these components work as electrodes to extract separated charge carriers from the absorbing material, while in the case of organic light emitting diodes (OLEDs), they inject charge carriers without affecting the light out-coupling efficiency (Fig. 1.1). TEs pervade modern technologies, providing a critical component of optical displays, solar cells, lasers, optical communication devices, and solid-state lighting [3–5].

In a material, the electrical conductivity is generally described through the imaginary part of the refractive index or the extinction coefficient [6]. In this regard, materials like metals that are highly conductive will not normally transmit visible light, while highly transparent media like oxide glasses behave as insulators. The challenge is to decouple these two properties in such a way that one can tune the electrical property of a material without affecting its optical properties and vice versa. To achieve this, until now, the majority of the effort has been in developing dielectric transparent media into conductive ones without significantly altering their transparency, such is the case of transparent conductive oxides (TCOs) [7–9]. Less effort has been made to increase the optical transparency of an electrically conductive material, which is the main focus of this thesis.

Besides optical transparency and electrical conductivity, other TEs' properties are crucial to ensure device performance, with specifications depending on the types of device. For example, while surface roughness and work function are insignificant for resistive touch screens, they are crucial for OLEDs and organic solar cells (OSCs) [10, 11]. Moreover, the optical haze of particular transparent electrodes is beneficial for high-efficiency solar cells but problematic for certain

D. S. Ghosh, *Ultrathin Metal Transparent Electrodes for the Optoelectronics Industry*,
Springer Theses, DOI: 10.1007/978-3-319-00348-1_1,

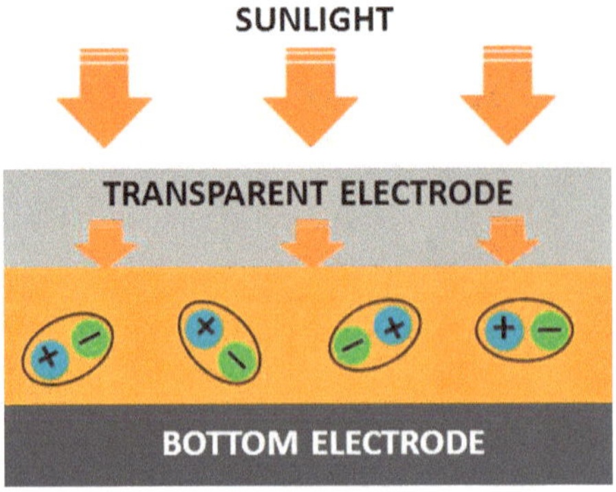

Fig. 1.1 Illustration of the role played by TE in a solar cell

types of displays [12]. The mechanical properties of the electrode can also be essential, e.g. for flexible devices. Harsh environment stability and impermeability to gas and water will be key properties investigated through this thesis.

1.1.1 Applications and Market

Over the last decade there has been a constant increase in devices which require one or more transparent conducting layers and that trend is expected to continue for the foreseeable future. Their ability to reflect thermal infrared (IR) heat is exploited to make energy conserving windows. Oven windows employ TEs to conserve energy and to maintain an outside temperature that makes them safe to touch. Automatically dimming rear view mirrors for automobiles and electrically controlled smart windows incorporate a pair of TEs with an electrochromic material between them. TEs can also be formed into transparent electromagnetic shields, invisible security circuits on windows, and transparent radio antennas built into automobile windows. The electrical conductivity of TEs is exploited in front surface electrodes for solar cells and flat panel displays which includes liquid crystal display (LCD), plasma and OLED based displays, larger and high resolution flat screens for portable computers. In fact, LCD is by far the largest user of TEs but many other devices are showing rapid growth such as touch screens, E-paper and flexible displays. Figure 1.2 shows the growth of TEs market in various applications. In particular, the total market value is expected to double over the next 4 years. With the advancement in thin film solar cell technology (roll to roll processing), it is anticipated that photovoltaic industry together with displays would be one of the largest consumer of TEs in the near future.

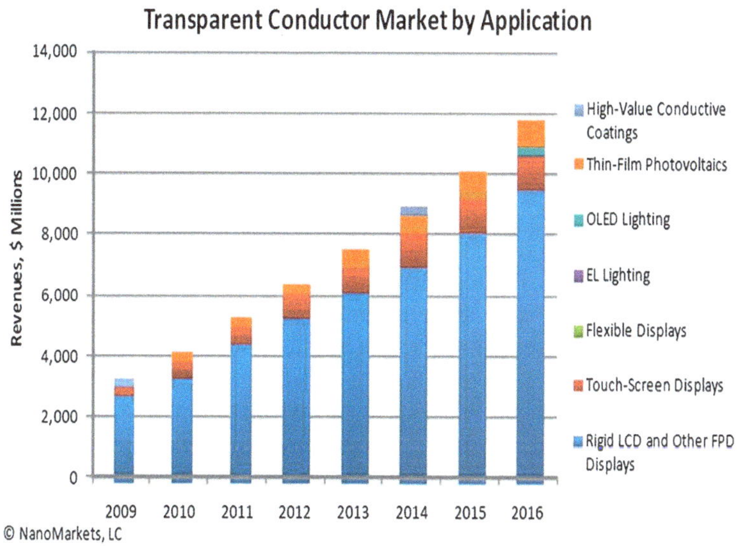

Fig. 1.2 Market value prediction of TEs for different applications (*Source* www.nanomarkets.com)

1.1.2 State of Art and its Drawbacks

The state-of-the-art solution relies on large band gap semiconductors doped with metal, which are also known as TCOs. Among TCOs, the mostly widely material used today is tin doped indium oxide (ITO). It has single-handedly dominated the field for almost four decades [13]. ITO has been the subject of research and has been refined for over 50 years and as a result, the material offers many beneficial properties that have made it the material of choice. The ability to deposit it with precisely controlled thicknesses and controlled doping concentration has significantly contributed to its widespread application [14]. A typical ITO film (100 nm) film has sheet resistance (R_S) of about 15–20 Ω/sq and average visible transparency >85 %, a trade off that meets most of the application-specific device requirements.

However, ITO has certain drawbacks, mainly centered on its scarcity of supply. It is widely believed that a shortage of indium may occur in the near future because of the limited nature of world indium reserves. The price of indium has recently increased by approximately one order of magnitude [15, 16]. As a result, a stable supply of ITO may be difficult to achieve considering the expansion of the TEs' market. Figure 1.3 shows the typical distribution of cost per m^2 for the fabrication of ITO film. As can be seen, a major chunk of the cost (>40 %) is the sputtering target. Apart from the supply and cost factor, the next generation of optoelectronic devices requires TEs to be flexible and compatible with large scale manufacturing methods (roll to roll processing). ITO, being an oxide, is brittle and therefore cracks when bended which leads to loss of its electrical properties [17]. Another

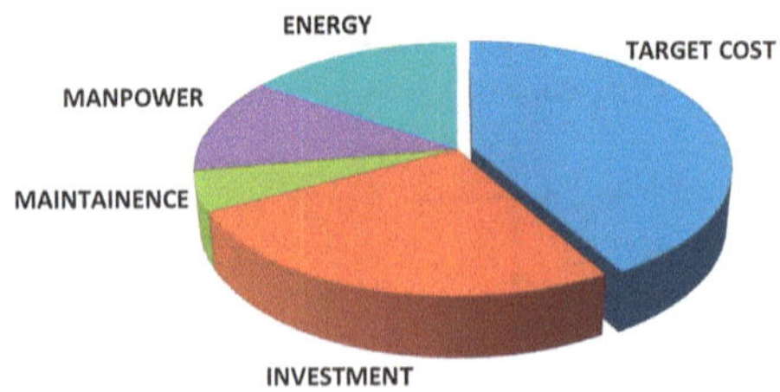

Fig. 1.3 Price distribution per m^2 for ITO film via sputtering (*Source* Von Ardenne)

drawback of ITO is the fact that deposition of ITO films is far from being a straightforward process, since their electrical and optical properties depend on dopant concentration, defects, and vacancies and usually require high temperature post-deposition treatment, the latter often being not compatible with the other processes used to make the device. ITO can also lead to device degradation owing to indium and/or oxygen migration from it into active organic layers [18]. Moreover, the ITO work function strongly depends on the cleaning procedure which is very crucial for certain applications [19]. Last but not least, several optoelectronic devices such as ultraviolet (UV) photodiodes, UV LEDs, solar cells for space applications, or IR pyroelectric detectors require electrodes with high transparency in the UV and/or IR regions. The typical band gap energy, $E_g = 3.75$ eV (331 nm), and plasma resonances in the near IR [20] make ITO impractical as transparent electrodes in the UV and IR respectively, even when the free-electron density is changed significantly [21].

1.1.3 Materials for ITO Replacement

1.1.3.1 Alternative TCOs

There is a great diversity of alternative TCOs to replace ITO. These include aluminum doped zinc oxide (AZO), fluorine doped tin oxide (FTO), magnesium doped zinc oxide (MZO), antimony doped tin oxide (ATO), gallium doped zinc oxide (GZO), aluminum doped MZO (AMZO) and gallium doped MZO (GMZO). Among these, AZO and FTO are considered the most viable alternatives to replace ITO [22, 23]. AZO is inexpensive, relatively abundant, easy to prepare, and non-toxic. AZO, fabricated by sputtering, has already established itself and has over 90 % of the market for the top electrode of CIGS solar cells [24]. One of the major concerns for AZO is the electrical stability of the film in humid environments, especially for films thinner than 100 nm (explained in detail in Chap. 7). FTO is well suited for chemical vapor deposition (CVD) at high temperatures, thus allowing its deposition during the drawing of the glass substrate. However FTO does not allow an easy, accurate etch, limiting its applications [25]. In general, all other TCOs possess significantly higher resistivity compared to ITO and more importantly still suffer from the problem of not being mechanically flexible.

1.1.3.2 Metallic Nanowire Networks

Random networks of metallic nanowires (NWs) have recently been put forward to replace ITO. Metal nanowires carry the advantages of low R_S, high transparency and mechanical flexibility. The solution process is also a low cost manufacturing technique which is compatible with roll-to-roll techniques [26, 27]. Silver NWs can achieve R_S values of 13 Ω/sq and transparency of 85 %[28]. More recently, Cu NWs were also synthesized to overcome the Ag price. Latest results of Cu NWs based TE show R_S of 20 Ω/sq and 60 % transparency [29]. The relatively large diameter of NWs (50–200 nm) causes significant surface roughness in the films, which is not desirable for applications with thin active layers. Additionally, the size and spacing of the NWs in the mesh lead to significant light scattering, a phenomenon known as haze, which is undesirable in many display applications [8]. The long term electrical stability of such films is a concern, as silver can corrode through such reactions as oxidation or sulfidation, which will increase the electrical resistivity of the films, especially due to the formation of poor contact between NWs. Electromigration of metallic ions in response to an applied current can also pose a potential problem in devices in which a high current carrying capacity is required [30].

1.1.3.3 Carbon Nano Tubes (CNTs)

CNTs are microscopic long, thin cylinders of carbon. They were first discovered in 1990s, and have remarkable physical and electrical properties. One of the critical requirements for CNT films is that the density of nanotubes must be above the threshold for the formation of a percolation network. Although the conductivity of individual nanotubes is high, the high resistance at the nanotube–nanotube junction limits the conductivity of the films [31]. In general, CNT films are comparably inexpensive and can be fabricated over large areas in various thicknesses and patterns. They are highly flexible and therefore are very well suited for flexible electronics. However, despite their properties, CNTs have been slow to find a commercial use. One of the major impediments towards implementing CNTs is the inability to obtain bulk quantities of material with suitable purity. In addition, their performance is still behind that of ITO films. For example, the best trade-off between R_S and transparency, reported so far, has been 30 Ω/sq and 70 %, still far from the requirements of most applications.

1.1.3.4 Graphene

Graphene is a two-dimensional allotrope of carbon and is an emerging material of interest, both for fundamental understanding and potential applications. It is a single sheet of sp^2 bonded carbon atoms having bandgap of 0 eV and thickness 0.34 nm. Graphene is one of the most transparent materials which absorbs only

2.3 % of the incoming light over the whole spectrum and has a theoretical R_S value of about 30 Ω/sq. The typical experimental sheet resistance of undoped synthesized graphene is between 2 and 5 kΩ/sq, roughly 100 times larger than the theoretical minimum [32]. Combining four layers of graphene leads to a pn experiemntal R_S value of ~350 Ω/sq maintaining 90 % transparency [33]. Various doping methods have been suggested to make graphene more conductive. Although the transparent electrode's performance of the doped graphene is much better than undoped graphene, its stability is an issue, especially in air and under thermal loading [34]. Graphene is usually grown on a substrate different from that of the final substrate used for device fabrication. The current graphene transfer procedures rely on polymer coating to maintain the graphene film integrity and to prevent folding, while the underlying Cu foil is chemically etched. Unfortunately, these process need several wet chemical steps that contaminates and mechanically damage the graphene resulting in higher R_S [35].

1.1.3.5 Conducting Polymers

Conductive polymers are those that are doped to create semiconducting materials. They are the essential materials platform for both thin film transistors and organic LEDs. Among them, poly (3,4-ethylenedioxythiophe) (PEDOT) is the most commonly used. It has a high electrical conductivity, up to 600 S cm^{-1}, and electrochemical stability while still maintaining moderate transparency and good film-forming properties [36]. It is often doped with aqueous polystyrensulfonic acid (PSS), which makes it water soluble and thus the processing becomes easier. The films are non-brittle and being in solution state makes them highly versatile. The best PEDOT-PSS film can achieve R_S values of 150 Ω/sq with 80 % transparency. However, the main problem of conducting polymers is the instability of the doped state, which can rapidly decrease the electrical conductivity when exposed to various stresses (thermal, chemical, UV light, etc.) [37].

1.2 Aim of the Thesis

The main objective of this thesis has been to develop new geometries of ultrathin metal film (UTMF) based TEs to replace costly and problematic ITO.

Metal films, when sufficiently thin (<10 nm), become transparent to light, yet still maintain good electrical conductivity. Single or multilayer UTMFs can overcome the high cost of raw materials such as indium and can be grown using single a process technique, i.e., sputtering, thus becoming a straight forward solution, that can to be integrated into typical industrial processes. The cost of depositing UTMFs based on abundantly available inexpensive metals can be 60 % less than that of typical ITO films (Fig. 1.3). Contrary to TCOs, UTMF based TEs possess high compatibility with nearly all organic and semiconductor materials (e.g., active medium) and related fabrication steps.

Another objective of the thesis has been to integrate the developed UTMF based TEs into energy efficient photonic devices. Different UTMF based geometries, compositions, and combinations have been investigated for potential applications in OLEDs, PV cells—organic and inorganic—and displays.

In addition, the possibility of combining the developed UTMF technology with other ITO replacing contenders like AZO was also explored to achieve improved hybrid TEs.

Through collaborative research, OPV cells and OLEDs which show electro-optical performance (efficiency) comparable with devices based on ITO electrode has been also achieved.

1.3 Thesis Outline

This thesis is organized in seven chapters. Chapter 2 provides an introduction of UTMFs, their deposition and characterization techniques. In particular the optical transmission of UTMFs from UV to mid-IR region (175 nm–25 μm) is investigated and a method to make the films environmentally stable is proposed. Such a method consists of carrying out appropriate temperature cycles in the presence of oxygen to produce a protective oxide layer on top of the metallic film which prevents further oxidation of the combined structure.

We have proposed a novel metallic grid based structure and reported experimental results which confirm that it is possible to reduce the R_S of UTMF based TEs of more than two orders of magnitude without significantly affecting their transparency, as described in Chap. 3. The chapter also discusses the grid design for the optimum trade-off between optical and electrical properties.

Copper which is widely used in microelectronics has excellent electrical and optical properties. However it is very unstable and tends to form oxide in the presence of oxygen. Chapter 4 describes the development of bilayer TE structure consisting of an ultrathin and continuous Cu film covered by a protective ultrathin Ti or Ni film. The capping layer is chosen according to the functional requirements of specific devices. The performance of these Cu based TEs has been investigated in devices like OLEDs and OPVs, achieving similar efficiencies to those of ITO.

The visible optical spectrum of UTMFs within the visible range is non-uniform due to the presence of a characteristic plasma frequency and intraband electronic transitions. In Chap. 5, a detailed theoretical and experimental investigation on utilizing alloy films as TEs has been described, which can offer a flatter wavelength response with respect to single component UTMFs. Efficient OPV cells employing ultrathin alloy electrodes were also demonstrated.

In Chap. 6, a bilayer structure of metal/oxide (Al:ZnO capped Ag nano-thick film) has been described which can overcome the high reflection of metals yet retain their good electrical behavior, while still keeping the minimum total film-thickness. Electrical and optical properties as well as the environmental stability of the bilayer structure were theoretically and experimentally studied. In particular, the use of UTMFs to improve the stability of Al-doped ZnO (AZO), which is known to become a problem at small thicknesses, is proposed which is described in Chap. 7. In the same chapter, the proposed TEs were shown to be used efficiently in OLEDs and OPVs.

And, finally, in Chap. 8, we conclude with some of our ongoing effort in the field and future outlook.

References

1. D.S. Ginley, H. Hosono, D.C. Paine, *Handbook of transparent conductors* (Springer, Berlin, 2010)
2. A. Facchetti, T.J. Marks, *Transparent electronics* (Wiley, Chichester, 2010)
3. C.G. Granqvist, Transparent conductors as solar energy materials: A panoramic review. Sol. Energy Mater. Sol. Cells **91**, 1529 (2007)
4. R.G. Gordon, Criteria for choosing transparent conductors. MRS Bull. **25**, 52 (2000)
5. J.F. Wager, D.A. Keszler, R.E. Presley, *Transparent electronics* (Springer, Berlin, 2010)
6. G.J. Exarhos, X.D. Zhou, Discovery-based design of transparent conducting oxide films. Thin Solid Films **515**, 7025 (2007)
7. D.S. Ginley, C. Bright, Transparent conducting oxides. MRS Bull. **25**, 15 (2000)
8. B.G. Lewis, D.C. Paine, Applications and processing of transparent conducting oxides. MRS Bull. **25**, 22 (2000)
9. T. Minami, New n-type transparent conducting oxides. MRS Bull. **25**, 38 (2000)
10. L. Hu, H. Wu, Y. Cui, Metal nanogrids, nanowires, and nanofibers for transparent electrodes. MRS Bull. **36**, 760 (2011)
11. S.K. Hau, H.L. Yip, A.K.Y. Jen, A review on the development of the inverted polymer solar cell architecture. Polym. Rev. **50**, 474 (2010)
12. Y. Nasuno, N. Kohama, K. Nishimura, T. Hayakawa, H. Taniguchi, M. Shimizu, Effect of perforated transparent electrodes on light transmittance and light scattering in substrates used for microcrystalline silicon thin-film solar cells. Appl. Phys. Lett. **88**, 71909 (2006)
13. D.S. Hecht, L. Liangbing, G. Irvin, Emerging transparent electrodes based on thin films of carbon nanotubes, graphene and metallic nanostructures. Adv. Mater. **23**, 1 (2011)
14. A. Kumar, C. Zhou, The race to replace tin-doped indium oxide: which material will win? ACS Nano **4**, 11 (2010)
15. T. Minami, Present status of transparent conducting oxide thin film development for indium tin oxide substitutes. Thin Solid Films **516**, 5822 (2008)
16. C.A. DiFrancesco, M.W. George, J.F. Carlin Jr, A.C. Tolcin, USGS Indium Report (2007)
17. O. Inganas, Avoiding indium. Nat. Photonics **5**, 201 (2011)
18. J.C. Scott, J.H. Kaufman, P.J. Brock, R.D. Pietro, J. Salem, J. Goitia, Degradation and failure of MEH-PPV light-emitting diodes. J. Appl. Phys. **79**, 2745 (1996)
19. A. Andersson, N. Johansson, P. Broms, N. Yu, D. Lupo, W.R. Salaneck, Fluorine tin oxide as an alternative to indium tin oxide in polymer LEDs. Adv. Mater. **10**, 859 (1999)
20. I. Hamberg, C.G. Granqvist, Evaporated Sn-doped In2O3 films: Basic optical properties and applications to energy-efficient windows. J. Appl. Phys. **60**, R123 (1986)
21. C.G. Granqvist, A. Hultaker, Transparent and conducting ITO films: new developments and applications. Thin Solid Films **411**, 1 (2002)
22. B.H. Liao, C.C. Kuo, P.J. Chen, C.C. Lee, Fluorine-doped tin oxide films grown by pulsed direct current magnetron sputtering with an Sn target. Appl. Opt. **50**, C106 (2011)

23. S. Fern andez, F. B. Naranjo, Optimization of aluminum-doped zinc oxide films deposited at low temperature by radio-frequency sputtering on flexible substrates for solar cell applications, Sol. Energy Mater. Sol. Cells **94**, 157 (2010)
24. C.H. Huang, H.L. Cheng, W.E. Chang, M.S. Wong, Comprehensive characterization of DC sputtered AZO films for CIGS photovoltaics. J. Electrochem. Soc. **158**, H510 (2011)
25. H.L. Hartnagal, A.L. Dawar, A.K. Jain, C. Jagadish, *Semiconducting transparent thin films* (Inst. Phys. Publ., Philadelphia, 1995)
26. J.Y. Lee, S.T. Connor, Y. Cui, P. Peumans, Solution processed metal nanowire mesh transparent electrodes. Nano Lett. **8**, 689 (2008)
27. C.H. Liu, X. Yu, Silver nanowire-based transparent, flexible, and conductive thin film. Nano Express **6**, 75 (2011)
28. S. De, T.M. Higgins, P.E. Lyons, E.M. Doherty, P.N. Nirmalraj, W.J. Blau, J.J. Boland, J.N. Coleman, Silver nanowire networks as flexible, transparent, conducting films: extremely high DC to optical conductivity ratios. ACS Nano **3**, 1767 (2009)
29. A.R. Rathmell, S.M. Bergin, Y.L. Hua, Z.Y. Li, B.J. Wiley, The growth mechanism of copper nanowires and their properties in flexible, transparent conducting films **22**, 3558 (2010)
30. J.M. Baik, S.J. Lee, M. Moskovits, Polarized surface-enhanced Raman spectroscopy from molecules adsorbed in nano-gaps produced by electromigration in silver nanowires. Nano Lett. **9**, 672 (2009)
31. Z. Li, Y. Jia, J. Wei, K. Wang, Q. Shu, X. Gui, H. Zhu, A. Cao, D. Wu, Large area, highly transparent carbon nanotube spiderwebs for energy harvesting. J. Mater. Chem. **20**, 7236 (2010)
32. F. Bonaccorso, Z. Sun, T. Hasan, A.C. Ferrari, Graphene photonics and photovoltaics. Nat. Photonics **4**, 611 (2010)
33. X. Li, Y. Zhu, W. Cai, M. Borysiak, B. Han, D. Chen, R.D. Piner, L. Colombo, R.S. Ruoff, Transfer of large area graphene films for high performance transparent conductive electrodes. Nano Lett. **9**, 4359 (2009)
34. R. Jackson, B. Domercq, R. Jain, B. Kippelen, S. Graham, Stability of doped transparent carbon nanotube electrodes. Adv. Funct. Mater. **18**, 2548 (2008)
35. X. Liang, B.A. Sperling, I. Calizo, G. Cheng, C.A. Hacker, Q. Zhang, Y. Obeng, K. Yan, H. Peng, Q. Li, X. Zhu, H. Yuan, A.R.H. Walker, Z. Liu, L.M. Peng, C.A. Richter, Toward clean and crackless transfer of graphene. ACS Nano **5**, 9144 (2011)
36. S.I. Na, S.S. Kim, J. Jo, D.Y. Kim, Efficient and flexible ITO-free organic solar cells using highly conductive polymer anodes. Adv. Mater. **20**, 4061 (2008)
37. S. Sakkopoulos, E. Vitoratos, E. Dalas, Conductivity degradation due to thermal aging in conducting polyaniline and polypyrrole. Synth. Met. **92**, 63 (1998)

Chapter 2
Basics of Ultrathin Metal Films and Their Use as Transparent Electrodes

2.1 Definition of Ultrathin Metal Films

The beginning of "Thin Film Science" can possibly be traced back to the observation of Grove [1] in 1852 who noted that metal films are formed by sputtering of cathodes with high energy positive ions. Since then it has come a long way and today it has become a fully-fledged academic discipline which has led to many industrial and household products. There is a phenomenal rise in thin metal films research, like their counterpart dielectric films, due to their extensive applications in electronics, optics, aviation, space science, defence and several other industries [2, 3]. These investigations have led to numerous inventions in the form of active devices and passive components such as piezo-electric devices, sensor elements, storage of solar energy and its conversion to other forms, reflecting and anti-reflecting coatings and many others. Furthermore, due to compactness, better performance and reliability coupled with low cost production, thin film devices and components are preferred over their bulk counterparts [4, 5].

Metal film properties are sensitive not only to their structures but also to many other parameters, including the process by which they are made, deposition conditions, substrates onto which they are deposited and most crucially on their thickness [6–8]. Basic research on thin metal films is generally limited to certain ranges of thickness, typically between a few Angstroms (Å) to about 10,000 Å depending on the properties under investigation. Mathematically, thin films are defined as a homogeneous solid material contained between two parallel planes and extended infinitely in two directions (x, y) but restricted along the third direction (z), which is perpendicular to the x–y plane. The dimension along the z-direction is known as the film thickness. Within the definition of thin films they can be classified according to their thickness as (i) ultrathin, (ii) thin, and (iii) comparatively thicker ones, the last one generally greater than 1000 Å. The first category has thickness in the range of few Å to 100 Å while the second category falls in the regime of 100–1000 Å. UTMFs ($\leq$10 nm) find their applications in areas such as magnetic sensors, recording materials, electro-optic and novel

D. S. Ghosh, *Ultrathin Metal Transparent Electrodes for the Optoelectronics Industry*, Springer Theses, DOI: 10.1007/978-3-319-00348-1_2,

devices such as spin filters, transistors, solar cells, and organic LEDs. In this thesis work as mentioned in Chap. 1, the UTMFs were explored and successfully used as TEs which can seriously compete with widely used indium based ITO.

2.2 Deposition of UTMFs

The existing techniques for thin film deposition are essentially based on Physical Vapor Deposition (PVD) processes: among those, the most popular ones are sputtering and thermal evaporation, but also pulsed laser ablation is gaining importance. Whichever the technique used, due to the importance of impurities and contaminants, deposition of thin films using PVD relies on ultra-high vacuum systems, i.e. base vacuum levels of $>10^{-7}$ Torr [9, 10]. Moreover, the nature of the deposited films is governed by a great variety of parameters that characterize the deposition process, such as deposition rate, pressure and composition of deposition atmosphere, cleanliness level of the vacuum chamber, substrate temperature, substrate materials, target-substrate distance, and substrate angle [8–10]. The presence of defects, dislocations, grain boundaries and surface roughness may be determinant for the film properties. In order to reach film thicknesses below 100 Å (10 nm), one must often push process parameters (such as vacuum base-level, gas pressure, power, time of exposure) to the extreme limits allowed by the deposition system themselves. This may lead to problems of film quality (for example in terms of uniformity) and repeatability. These drawbacks are critical, and have to be taken into account when dealing with industrial manufacturing.

Among the variety of PVD processes, deposition by sputtering holds a unique position. Well known advantages over evaporation processes include wide-area uniform coverage, ability to coat temperature-sensitive substrate materials, and the virtually limitless variety of materials that can be deposited [11]. Included in the latter category are metals, compounds, alloys, and mixtures. Oxide, nitride, and fluoride compounds can be reactively sputtered starting from metal targets. Many variants of sputtering exist today, with DC sputtering the most appropriate for metals while RF sputtering for dielectrics and special compositions.

The basic process of sputtering involves a process gas (in our case argon, Ar) which is introduced inside the high vacuum chamber (Fig. 2.1). An electric field is applied between the target and the substrate which leads to the creation of plasma with the positive Ar^+ ions accelerated towards the cathode (target) located on the bottom of the chamber. Particles are detached from the target due to the momentum transfer of Ar^+ ions when colliding with the target material, contained within a plasma plume similar to a flash light or a cone resulting in deposition of film on the substrate situated just above the cathode. Free electrons generated during the process are confined close to the target material using a magnetic field created by a magnet ring located just behind the cathode (target). It is then less probable that these ions recombine with free electrons in their way to the target, leading to an increased sputtering rate.

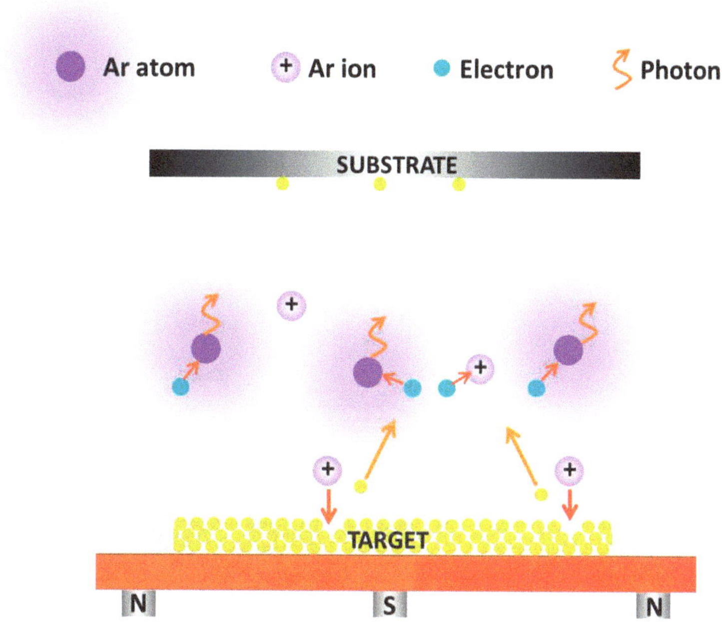

Fig. 2.1 Spatial distribution of particles inside the high vacuum sputtering chamber

In ICFO, we have two different kinds of sputtering system for deposition of UTMFs: ATC Orion 3 HV and, a more recent, advanced ATC Orion 8 HV. Both of them are magnetron sputtering systems having different configurations and are manufactured by AJA International Incorporation [12]. Orion 3 HV is a single target, low throughput R&D engine, allowing both DC and RF magnetron sputtering in pure Ar atmosphere. Its set-up is planar, with a fixed target (and magnetic field) and a pendulum oscillating platform for the substrate. No degas chamber is available and depositions can only be made at room temperature. Orion 3 HV's pumping system is composed of dry rotary vane pump which allows reaching a base vacuum pressure level of 10^{-7} Torr (4 h stand-by) in the deposition chamber. Each target has a diameter of 2 in. allowing good deposition uniformity (<2.5 %) over a maximum area of ~4 in. The ATC Orion 8 HV sputtering system instead is a computer controlled fully automatic RF/DC deposition system with co-planar configuration and can have 7 different target materials installed at the same time. The target size diameter is 2 in. and the system reaches thickness uniformity of 2.5 % over 4 in. diameter substrates. It has an integrated load lock system for sample transfer without breaking the vacuum of the main chamber. The ATC Orion 8 HV has 2 radio-frequency (RF) and 2 direct current (DC) power sources which allows co-sputtering as well. The system also allows deposition at higher temperature, up to 800 °C and has an O_2 reactive gas line, apart from an Ar line (process gas). Substrates are placed on a rotating sample holder that can spin around the axis of the chamber up to a maximum rotation frequency of 40 revolutions/min. The main chamber is connected to a turbo pump which can reach a base vacuum level of 5×10^{-8} Torr in just 3 h of standby time while the load lock is pumped by a smaller rotary pump. The system is also fitted with a quartz crystal thickness monitoring unit by which the deposition rates can be deduced (Fig. 2.2).

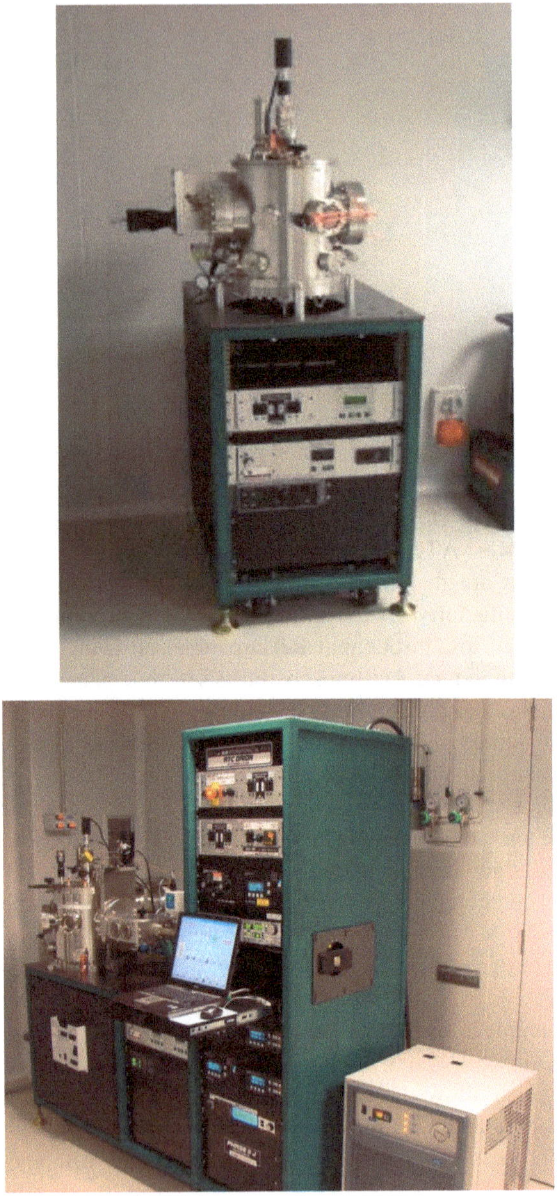

Fig. 2.2 Picture of ATC Orion 3 HV (*top*) and ATC Orion 8 HV (*bottom*) installed at Nanophotonics lab, ICFO

Several substrates can be used for UTMF deposition according to the different experimental requirements. Electrical resistivity measurements were carried out using non-conductive substrates while transparent substrates were used for optical transmittance studies. On the contrary, experiments, such as thickness measurements, do not have any special substrate requirement and silicon substrates are thus used for this purpose while for analyzing the flexibility of the UTMFs, polyethylene terephthalate (PET) of 0.125 μm thickness were used. For device applications, such as OPVs and OLEDs, double side optically polished UV fused silica substrates were preferred. These substrates are characterized by their very low surface roughness and are transparent in the UV region of the spectrum. In order to get rid of any possible environmental contaminant a standard cleaning technique prior to the deposition has been employed for all the samples. An ultrasonic bath in acetone (10 min) followed by ethanol (10 min) was employed. The samples were then dried using a N_2 flux. The cleaned substrates were subsequently cleaned again inside the main sputtering chamber for 15 min by using either an Ar or O_2 plasma.

2.3 Growth of UTMFs

Thin film growth consists of three distinctive steps: nucleation followed by coalescence and finally thickness growth. During the early stages of thin film formation a sufficient number of vapor atoms or molecules condense and establish a permanent residence on the substrate. Many such film birth events occur in this so called nucleation stage. In the nucleation stage, a uniform distribution of small but highly mobile clusters or islands is observed. In this stage the prior nuclei incorporate impinging atoms or molecules and subcritical clusters and grow in size while the island density rapidly saturates. The next stage involves merging of the islands by a coalescence phenomenon. Coalescence decreases the island density resulting in local denuding of the substrate where further nucleation can occur. Continued coalescence results in the development of a connected network with unfilled channels in between. With further deposition, the channels fill in and shrink leaving isolated voids behind. Finally, even the voids fill in completely and the film is said to be continuous.

Generally, during the nucleation stage depending on the interaction energies of the substrate atoms and the film atoms, three growth modes are observed: (i) Frank-Van der Merwe (layer) [13], (ii) Volmer-Weber (island) [14], and (iii) Stranski–Krastanov (mixed) [15]. In 1958, Bauer gave a thermodynamic criterion for the growth mode [16]. The criteria states that under equilibrium conditions, the growth mode is determined by the following energy difference:

$$\Delta\sigma = \sigma_f + \sigma_i - \sigma_s \tag{2.1}$$

where σ_s is the surface free energy of the substrate, σ_i is the free energy of the interface, and σ_f the free energy of the film (Fig. 2.3).

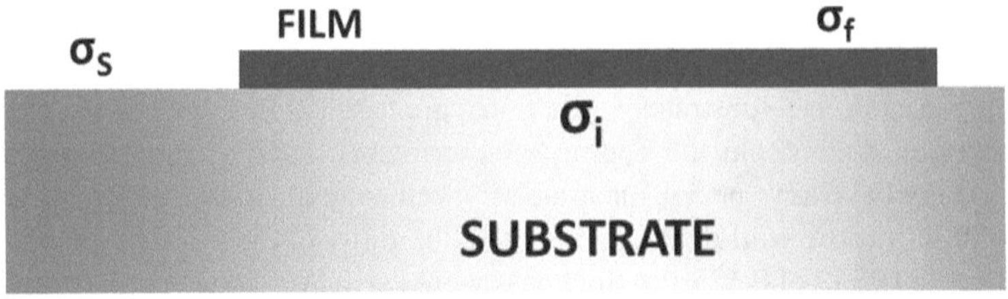

Fig. 2.3 Bauer's thermodynamic criterion for the growth mode

If $\Delta\sigma \leq 0$, the adsorbed atoms/molecules (adatoms) are more strongly bound to the substrate than to each other and the extension of the smallest stable nucleus occurs overwhelmingly in two dimensions, resulting in the formation of planar sheets. Under this condition, complete wetting of the substrate is favorable, and Frank–van der Merwe (layer-by-layer) growth should be observed. The inequality has the opposite sign when the atoms (or molecules) are more strongly bound to each other than to the substrate. In this case, one usually obtains Volmer–Weber growth (island growth), i.e., no wetting of the substrate. To predict the growth mode of UTMFs, one can look at results on metals and semiconductors deposited on oxide substrates. The film energy may have a contribution such as the strain energy, which increases linearly with increasing film thickness. After this energy is added to $\Delta\sigma$, it is possible for $\Delta\sigma$ to be smaller than zero until a certain thickness is reached and then larger than zero above this coverage. In this case, Stranski–Krastanov growth (mixed) generally occurs.

In our case, it is desired to have the percolation thickness (thickness at which the films becomes continuous) of metal films to be as low as possible. Generally, in case of metal films the adatoms are more bound to each other compared to their interaction with the substrates i.e. $\Delta\sigma \geq 0$ and therefore they grow via Volmer–Weber (island) mode. To increase the interaction energy with the substrates, as mentioned in Sect. 2.2, the substrates themselves were treated in O_2 or Ar gas plasma inside the sputtering chamber for 15 min.

2.4 Surface Analysis of UTMFs

Analyzing the surface morphology of UTMFs provides information on the structural properties of films as well as insights on electrical and optical performance. When passing from bulk to thin films, surface structures become more important in determining UTMF's properties. In particular, surface morphology dramatically affects electrical conduction. Non-conductive films might arise from sufficiently rough surfaces as a result of discontinuity. Accordingly, it is extremely important when dealing with UTMFs that the deposited films remain smooth when compared to the thickness of the layer. It is already intuitive that the surface roughness of the substrates has to be smaller than the thickness of the UTMF to be deposited if the latter is to be continuous [17]. As mentioned earlier, we used optically polished UV fused silica substrates for UTMF deposition which are characterized by a very low surface roughness (RMS roughness: <2 Å).

We used *Veeco Digital Instrument Dimension 3100* atomic force microscope (AFM) instrument for surface roughness measurements at atomic scale. The limited scan area (usually 10 × 10 μm) leads us to take a random area as a representative measurement for the rest of the surface. Typical 3D and profile images obtained with the AFM for Ni 5 nm and Cr 5 nm on UV fused silica substrates are shown in Fig. 2.4. Although some fluctuations larger than the thickness can be observed—suggesting that the films might have local discontinuities—the layers were found to be globally continuous as was also confirmed by electrical resistivity measurements. The raw images obtained using AFM techniques were treated using the associated software *Nanoscope Analysis*. The surface roughness is usually represented by RMS roughness (R_Q) which is given according to the following formula:

$$R_Q = \sqrt{\frac{1}{L}\int_0^L z^2(x)dx} \tag{2.2}$$

where '*L*' is the evaluation length, '*z*' is height and '*x*' is the distance along measurement.

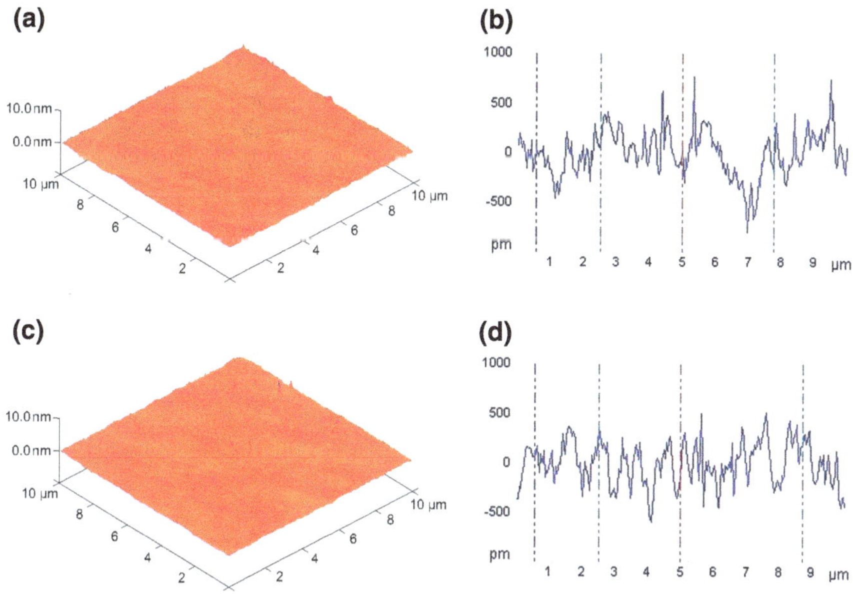

Fig. 2.4 Three dimensional and thickness profile images of Ni 5 nm (**a**, **b**) and Cr 5 nm (**c**, **d**) films on UV fused silica substrate. The RMS roughness of the films was to be 0.235 and 0.245 nm for Ni and Cr films respectively

2.5 Flexibility Analysis of UTMF

Analyzing the properties of UTMFs under stress provides information about the degree of robustness of the films. With the technology now moving toward flexible electronics with flexible OLEDs, OPV and displays, it is very important for the TEs to maintain their properties when bent. Changes in properties of TE would lead to a device with a shorter lifetime or to complete failure. To study the flexibility properties of deposited UTMFs, a set-up as shown in Fig. 2.5 was developed. The film to be analyzed is deposited onto 1 × 1 in. flexible substrate (generally polyethylene terephthalate (PET) or polyethylene-naphthalate (PEN) and is mounted onto the clips with two of its ends fixed. The set-up is connected to a motor which in-turn is controlled by an electronic controller which moves the arm in the horizontal direction. The cycling movement of the arm results in bending of the sample and the displacement is controlled in such a way that it makes a minimum radius of curvature of 4 mm and maximum of 12 mm on the flexible sample. The R_S value and transparency of the UTMFs is measured before and after a finite number of bending cycles to assess their flexibility and mechanical robustness.

2.6 Electrical Analysis of UTMFs

The characterization of electrical conduction of UTMFs is described in this section. By definition, intrinsic properties such as electrical resistivity or conductivity, should be constant and independent of size for a given bulk material. However, this is not true for thin films. When the film thickness becomes comparable with the characteristic length scale of a physical phenomenon, the intrinsic properties are affected by so called size-effects [18, 19]. The electrical conductivity is affected by the limitation of size as soon as film thickness becomes comparable to the electron mean free path in the medium. In the case of UTMFs, its electrical resistivity is found to be much higher than that of the corresponding bulk metal and it decreases with the increasing film thickness, eventually attaining a value often approaching that of the bulk. Before presenting the experimental results, a brief overview of the basic theory of conduction of thin metal films in the presence of enhanced size-effect is presented. The theoretical modeling of thin film resistivity is necessary for correct interpretation of experimental data.

2.6.1 Electrical Conductivity Mechanisms of UTMFs

Contrary to bulk metal, thin metallic films have one dimension, i.e. the thickness (t), which is much smaller than the other two dimensions in the plane of the film and is

Fig. 2.5 Set-up showing the bending apparatus

comparable, in not smaller, to the electron mean free path (MFP) in bulk (l_b). In this regime, the electrical resistivity depends on the thickness of the film. Thompson considered two conditions for the calculation of thin film conductivity: (i) diffuse scattering of conduction electrons (where scattering is independent of scattering angle) by the two film surfaces, and (ii) scattering by lattice matrices where MFP of the film (l_f) will be the same as that of the bulk l_b [20]. This can be seen in Fig. 2.6.

For some arbitrary starting point 'z' within the film, electrons will be scattered in all directions. However, only in the region defined by ϕ_2–ϕ_1, the electrons will be able to travel up to their MFP. While in other regions, the electrons will strike the surface of the film. From these considerations, Thomson calculated the following relations

$$l_f = \frac{3}{4}t + \frac{1}{2}t\,\log\left(\frac{l_b}{t}\right) \tag{2.3}$$

$$\frac{\sigma_f}{\sigma_b} = \frac{l_f}{l_b} = \frac{3t}{4l_b} + \frac{t}{2l_b}\log\left(\frac{l_b}{t}\right) \tag{2.4}$$

$$= \frac{3k}{4} + \frac{k}{2}\log\frac{1}{k} \tag{2.5}$$

where subscripts 'f' and 'b' stand for film and bulk, respectively, and $k = t/l_b$. Thompson's approach suffers from the fact that it neglects free paths which start from the film surface.

Fuchs [21] and Sondheimer [22] (known as FS model) made a rigorous analysis of the electrical conduction process in an ideal continuous metallic film bound by two plane surfaces assuming the Boltzmann equation for the distribution function

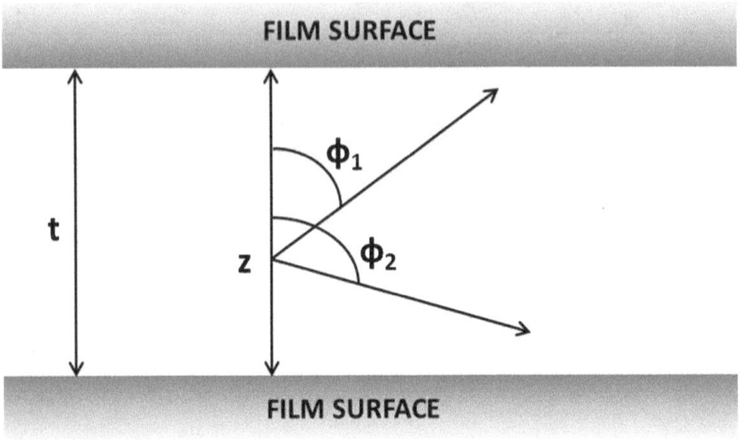

Fig. 2.6 Diffuse scattering of electrons by a film surface leading to thickness dependent mean free path l_f

and also considering the boundary scattering process. The relation between two conductivities i.e. bulk (σ_b) and film (σ_f) is then given by

$$\frac{\sigma_b}{\sigma_f} = \frac{\rho_f}{\rho_b} = \emptyset_k/k \tag{2.6}$$

where $1/\emptyset_k = (1/k) - \left(\frac{3}{8k^2}\right) + (3/2k^2)\int_1^{\infty}\left(\frac{1}{t^3} - \frac{1}{t^5}\right)e^{-kt}.dt.$

This equation can be further simplified for thin films, i.e. $t < l_b$ or $k < 1$ to

$$\frac{\rho_f}{\rho_b} = \frac{4}{3}\frac{1}{k\{ln(1/k) + 0.4228\}} \tag{2.7}$$

Sondheimer further assumed that if a fraction 'p' of electrons is specularly scattered from the two surfaces and the rest are diffused with complete loss of their velocity, then Eq. 2.7 assumes the form

$$\frac{\rho_f}{\rho_b} = 1 + \frac{3}{8k} \qquad \text{for k} \ll 1 \tag{2.8}$$

and

$$\frac{\rho_f}{\rho_b} = \frac{4}{3k}\frac{1+p}{1-p} \qquad \text{for k} \gg 1 \tag{2.9}$$

Although these typical theoretical models are adequate to explain the electrical behavior of thin metal films, they might not be very accurate in the case of UTMFs. For this reason, several new approaches have been considered during the

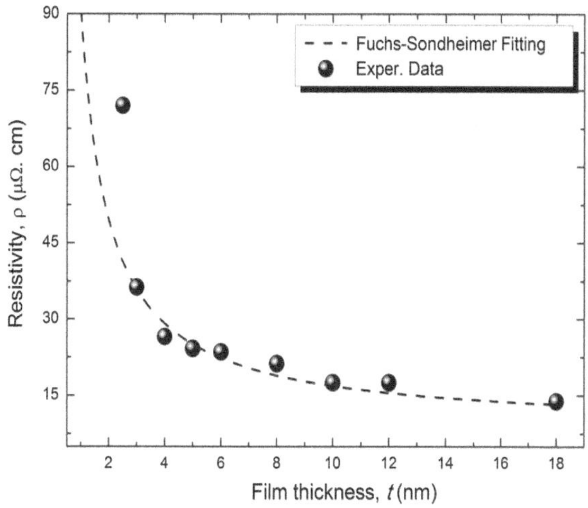

Fig. 2.7 Experimental data for Ni layers fitted with F–S model

last decades [23, 24]. However, the FS approximation still remains the most accurate model to predict the electrical behavior of UTMFs as depicted in Fig. 2.7.

Besides geometrical limitation that enhances surface scattering, there are other sources of resistivity:

- Volume defects and impurities
- Lattice vibrations
- Strain and discontinuities, including grain boundaries.

All these are "volume" sources of scattering. Therefore, they are not included in FS expressions. As such, they should be the same for all films grown in the same conditions. Our UTMFs are polycrystalline in nature which are composed by single crystal grains. The presence of grain boundaries may reduce significantly the conductivity of metallic materials. Mayadas and Schatzkes [20] were the first to provide a model that takes into account both the scattering at external surfaces (size-effect) and scattering due to grain boundaries. They showed how a reduction of the grain size can induce a significant increase of resistivity. On the other hand, voids may also play a role in determining the electrical behavior of UTMFs. For example, an island growth can result in a discontinuous film. Even though characterized by very high resistivity, discontinuous films may still be conductive. This is due to quantum mechanical tunneling between individual islands or thermoionic emission of electrons into the conduction band of the substrate [25]. UTMFs, which grow in a Volmer–Weber or island mode, present many voids between the islands which increase the resistivity by up to orders of magnitude compared to that of a uniform film (Frank–van der Merwe or layer-by-layer growth).

2.6.2 Sheet Resistance Measurement

As mentioned in the previous Sect. 2.6.1, the UTMF′s electrical resistivity depends on the thickness. It is very common to measure the electrical surface properties of a film though the Sheet Resistance (Rs) in Ω/sq units:

$$\rho_t = R_S \, . \, t \tag{2.10}$$

where ρ_t is the thin film electrical resistivity and t is the thickness. The sheet resistance can be measured directly using a Four-Point probe set-up connected to a multimeter. The schematic of the set-up is shown in Fig. 2.8.

To obtain the electrical resistivity of the UTMFs, the sheet resistance was measured using a *Cascade Microtech 44/7S 2791* Four Point Probe and a *Keithley 2001* multimeter. The four tips installed in the Four Point Probe were brought into contact with the layer. A test current has been injected through the outer tips while the inner tips collected the voltage drop as shown in Fig. 2.8. Typically 6 measurements were taken, each at different position on the film and the mean value of sheet resistance is calculated as:

$$R_S = C_1 \times C_2 \times R \tag{2.11}$$

where R is the average electrical resistance obtained by the multimeter and the two corrective coefficients C_1 and C_2 take into account the separation between tips (s) in the Four Point Probe Head and substrate's dimensions. In our case, the separation between tips in the four point probe head is about 1 mm and the diameter of the samples is 2.5 cm (1 in.). As UTMF thickness is always much lower than the spacing between tips, the first corrective coefficient is kept at a constant value ($C_1 = 1$) while the second one (C_2) is calculated to be 4:4364 for 1 in. substrates. We finally inferred the electrical resistivity of the layers using the definition of sheet resistance given by Eq. 2.11.

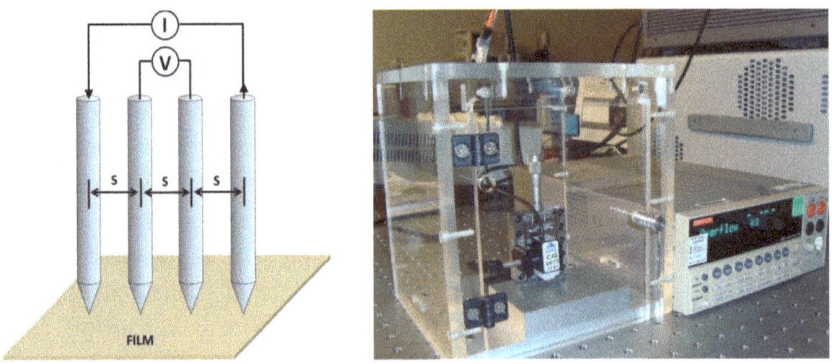

Fig. 2.8 Schematic of the Four Point probe set-up used for the sheet resistance measurement (*left*) and the actual set-up in ICFO (*right*)

2.7 Optical Analysis of UTMFs

In order to be competitive TEs for different applications, UTMFs should keep optical losses minimum in the visible region of the spectrum. In addition, there are applications that benefit from a very broad transparency spectrum, such as ultra-violet (UV) and infra-red (IR) detectors which work with wavelengths outside the visible range. In general the response of a material to optical radiation depends on its electronic band structure. In metals the optical properties depend mainly on the contribution of the conduction band, i.e. of the free electron gas, and in some cases on intra band transitions. As long as the relaxation time of free electrons is much shorter than the period of the electromagnetic wave, the metal will exhibit a strong absorption and a high reflectivity. Indeed the extinction coefficient k of metals has very high values at low frequencies while it becomes negligible for frequencies beyond the plasma frequency, which is generally in the UV or visible region.

Thin metal films, however, present reduced optical absorption and enhanced transmission due to the reduction of thickness. Optical absorption becomes negligible when the film thickness is less than the penetration depth at optical frequencies. This is usually achievable for thicknesses <10 nm. Optical properties of UTMFs are also affected by the size-effect, since the dielectric function of metals depends mainly on the conductivity and the number of free electrons. For example, a shift of the plasma frequency should be expected when the resistivity starts to increase due to surface scatterings. Moreover, intrinsic optical properties may further differ from those of the corresponding bulk, due to the polycrystalline nature of films. Grains or macroscopic surface roughness may induce the presence of voids and discontinuities inside the volume of the metal film, giving rise to a non-uniform distribution of matter (Fig. 2.9).

As it was mentioned above, several optoelectronic devices, such as ultraviolet (UV) photodiodes, UV LEDs, solar cells for space applications, and infrared (IR)

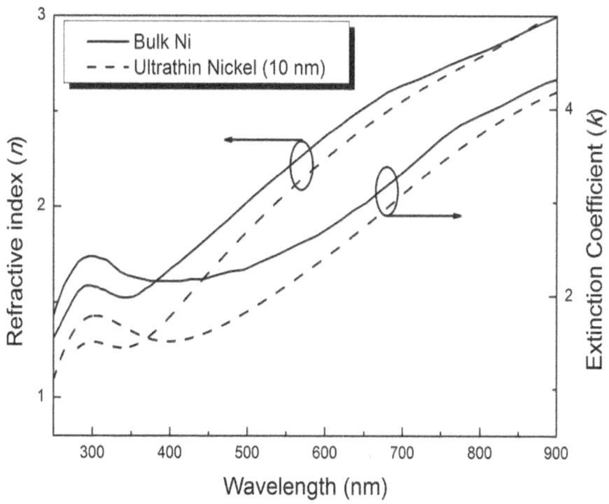

Fig. 2.9 Comparison of optical constants of bulk and 10 nm nickel film. The measurements on UTMFs are carried out using *Sopra GES 5E Ellipsometer* while bulk values are taken from Ref. [26]. UTMFs have lower refractive index and extinction coefficient compared to their bulk counterparts

pyroelectric detectors, require electrodes with a high transparency in the UV and/or IR regions. The most widely used transparent electrodes, in particular Indium tin oxide (ITO), possess a typical band gap energy of about, $E_g = 3.75$ eV (331 nm) and plasma resonances in the near IR which makes them impractical as transparent electrodes in the UV and IR. In fact, the transparency of ITO can be increased in the near IR spectrum by reducing the free-electron density which in turn increases its resistivity [27]. Single component nickel and chromium UTMFs were deposited on UV fused silica and silicon substrates to investigate their broad spectrum optical response. Thicknesses were inferred from deposition rates determined by the quartz crystal thickness monitoring unit. They were also compared to state-of-art ITO in terms of both electrical and optical properties.

For comparison, two different ITO were grown: 28 nm as-deposited and 28 nm annealed at 450 °C for 2 h in normal atmospheric conditions. In addition, standard ITO sample (100 nm deposited onto Corning glass) was also purchased for comparison. Transmittance spectra were taken with a PerkinElmer Lambda 950 spectrometer in the UV–Visible region whereas a Shimadzu FTIR-8400S Fourier transform IR spectrometer was used in the mid-IR range. Note that substrate's contribution is always taken into account in optical transmittance measurements as $T_f = T_t/T_s$, where T_t is the total optical transmittance (film and substrate), whereas T_f and T_s are, respectively, the film and substrate optical transmittance. Electrical resistivities of the samples were measured as explained in Sect. 2.6.2.

Figure 2.10 shows the average optical transmittance in the visible range (375–700 nm) and the electrical resistivity for the Cr and Ni films of different thicknesses together with those of ITO layers, both as-deposited and annealed. In the visible range the performance of UTMFs is comparable to that of ITO. In fact, Ni films present a similar optical transparency with a significantly lower electrical resistivity. However, the real advantage of UTMFs over ITO in terms of optical transmittance is in the UV and IR ranges.

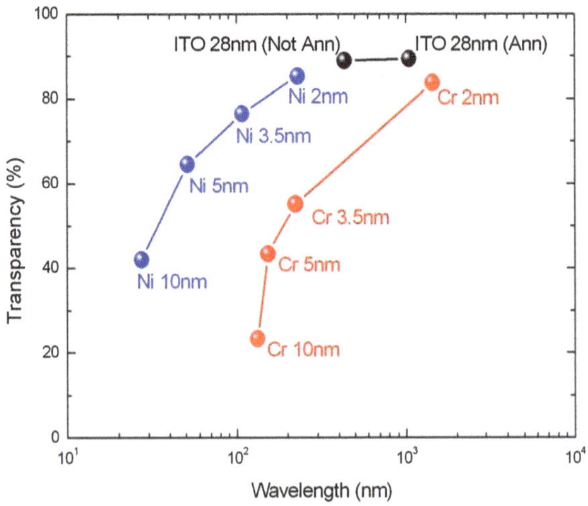

Fig. 2.10 Average optical transparency in the visible wavelengths against electrical resistivity for Cr and Ni films compared to ITO annealed and not annealed

In the UV range (175–400 nm), ITO films present enhanced absorption and therefore reduced transmittance since the material band gap is ~330 nm as discussed before. On the contrary, Cr and Ni films possess flatter optical transmittance with levels comparable to those in the visible range (Figs. 2.11).

Figures 2.12a and b show, respectively, the IR (from 2.5 to 25 μm) transmittance of Cr and Ni films, both being compared against ITO layers. As in the UV case, both Cr and Ni present superior optical transmittance properties: for thin samples both larger and flatter transmittance. We have also investigated the performance of standard ITO on corning glass, which presents a much faster decay of the transmission in the near IR; in fact, the transmittance goes below 20 % already at about 2.8 μm. The difference between the two ITO sets can be attributed to the

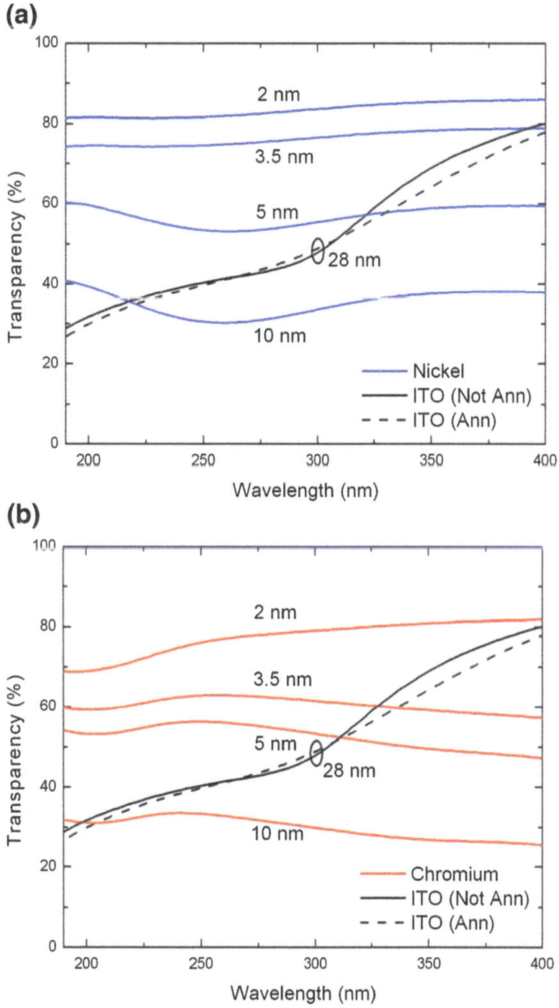

Fig. 2.11 Optical transmission of UTMFs compared to ITO annealed and as-deposited in the UV region: **a** Nickel and **b** Chromium

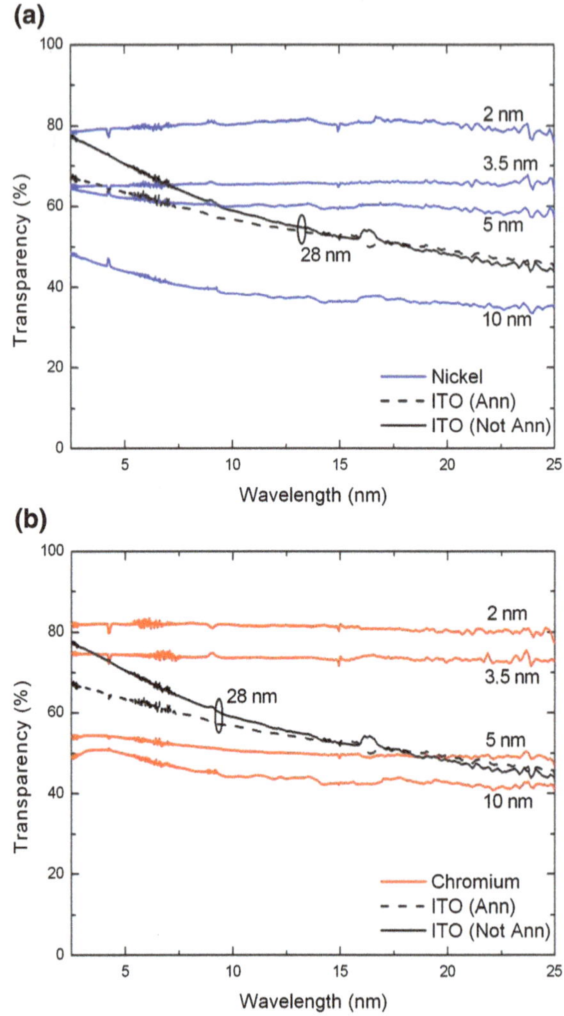

Fig. 2.12 Optical transmission of UTMFs compared to ITO annealed and not annealed in the mid-IR region: **a** Nickel and **b** Chromium

difference in free electron density [4]. The fact that the lower near-IR transparent ITO (on Corning glass) has a higher free-electron density than that of ITO on UV fused silica or Si is also confirmed by its lower electrical resistivity: about 160 μΩ x cm (the resistivity for ITO on UV fused silica is about 430 μΩ × cm).

2.8 Stability

A potential drawback of UTMFs for their use as transparent electrodes is the degradation they can undergo due to oxidation when exposed to environmental agents such as air, moisture and temperature. In fact, contact with other reactive chemical elements can also take place during fabrication (e.g. wet etching) or even be part of the structure of the device (e.g. metal/SiO2). UTMFs are thus always at risk of changing their electrical and optical properties. This aspect is also important for non-noble metallic layers, such as chromium (Cr), nickel (Ni), titanium (Ti) and aluminum (Al).

The effect of oxidation on the electrical and optical properties of UTMFs has been investigated while, on the other hand, we take advantage to induce *ad-hoc* oxide protective layers, so that stable UTMFs can be achieved. The work was focused on Ni films, which we subjected to thermal treatment in either ambient atmosphere ($\sim$20 % of O_2) or in the presence of a continuous O_2 flow. The obtained results and techniques can be extended to other metals, such as Cr, Ti and Al.

Several Ni UTMFs of different thicknesses were grown: 2.2, 3.4, 5, and 10 nm for the stability studies. The films were kept in ambient atmosphere for 12 days to analyze the environmental effect on the films. The corresponding relative variation in electrical resistivity ($\Delta\rho/\rho$) in % was then measured for the films. It was found that, for the thinnest sample (2.2 nm) film resistivity could not be measured any longer due to its high value while 3.4, 5 and 10 nm films showed increases of 59, 16 and 4 % respectively.

The samples were then subsequently thermally treated with increasing temperature, H1 to H4 (Fig. 2.13), carried out in ambient atmosphere using a Selecta High-temp oven. The temperature was measured using a Fluke thermometer 52 II conncctcd to an 80 PK-1 thcrmocouplc. Thc clcctrical rcsistivity kcpt incrcasing after each thermal treatment for samples 3.4 and 5 nm, this effect being larger for thinner films since the formation of a natural oxide layer due to oxygen indiffusion reduces the effective metallic path, thus leading to higher electrical resistivity. A final step (H5) identical to the first one (H1) was carried out at low temperature. As it can be observed in Fig. 2.13, contrary to the initial step (H1), the electrical resistivity changes due to H5 are negligible (within 2 %) for all films, thus indicating that all UTMFs have reached high stability.

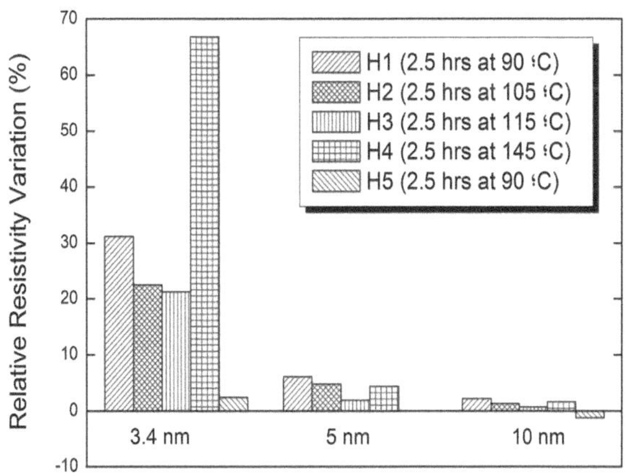

Fig. 2.13 Electrical resistivity variation after cumulative annealing treatment

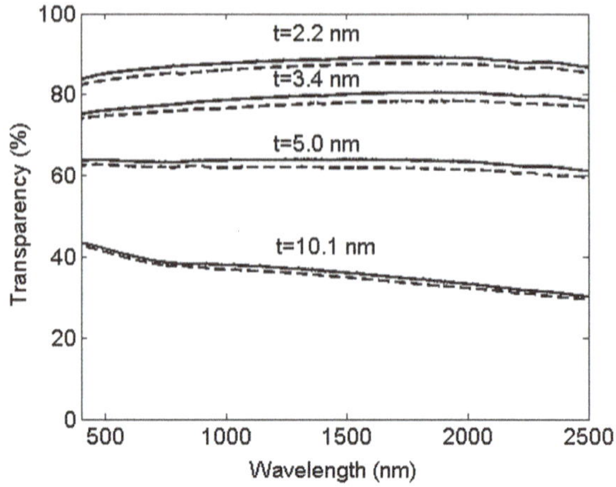

Fig. 2.14 Optical transparency of the films in the 400–2500 nm range (*dashed lines* represent as deposited samples while *solid lines* are obtained after step H5)

The optical transparency of all the samples was re-measured after the aforesaid thermal treatments and compared to that of as-deposited films (Fig. 2.14). All the samples showed increased transparency since metal oxides typically possess a lower extinction coefficient than the original metals, and the top oxide layer act as an index matching layer reducing the reflection and thus, increasing transmission [22].

In order to further elucidate the mechanism and increase the speed of the oxidation leading to stability, we subjected new Ni films to thermal treatment under a controlled dry O_2 flux. Films of different thickness were subjected to 90 min of degassing in vacuum at a temperature of about 140 °C and subsequently exposed to O_2 flux for 5 min by means of an ion gun (ion gun settings: 160 V acceleration potential and 1.8 Å ion current). The effect of oxidation can be estimated by evaluating the thickness variation of films, due to the formation of the corresponding metal oxide compound. Figure 2.15 shows the thickness scatter plot of different samples before and after the whole oxidation treatment. In case of

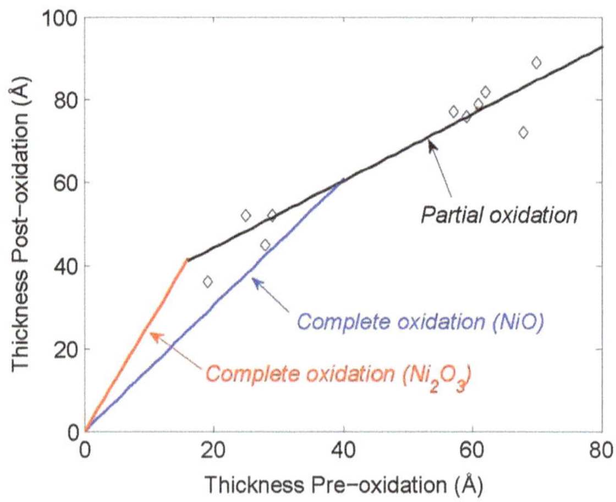

Fig. 2.15 Effect of oxidation on the thickness of nickel based UTMFs

Table 2.1 Molar mass and density of nickel and its main oxide compounds Ni (II) and Ni (III)

Element	Molar mass	Density	N/N_{ox}
Nickel	58.69	8.908	
NiO	74.71	7.45	1
Ni_2O_3	165.42	4.83	0.5

Last column gives the ratio between the compound number of moles over that of Ni

complete oxidation, and if the oxide compound is known, one can write the thickness ratio between metal and its oxide as:

$$\frac{t_{ox}}{t_m} = \frac{N_{ox}}{N_m}\frac{P_{ox}}{P_m}\frac{\delta_m}{\delta_{ox}} \tag{2.12}$$

where *N* is the number of moles, *P* is the molar mass and δ is the density of the material. Note that subscript *m* stands for metal and *ox* for the oxide. The main relevant parameters for Ni and its main oxide compounds are given in Table 2.1.

Given that Ni has two oxide compounds (NiO and Ni_2O_3), both corresponding complete oxidation lines are shown in Fig. 2.15 with slopes t_{ox}/t_m of 1.52 and 2.6, respectively. Completely oxidized films will increase their thickness according to a proportional factor within this interval, since the formation of a solid solution of both oxides might also be possible. However, in practice, oxidation is expected only up to a maximum thickness. Thicker layers should undergo partial oxidation for the same amount of metal thickness; in this case Eq. (2.12) will not hold. By fitting experimental data of thicker samples one obtains the partial oxidation line whose intersection point with the complete oxidation line represents the maximum oxidation thickness. According to thickness variations, the maximum oxidation thickness under the experimental conditions is in the range of 14–40 Å.

The electrical resistivity measurements of a passivated Ni 10 nm film were also carried out as a function of temperature in the 30–150 °C range (Fig. 2.16). The

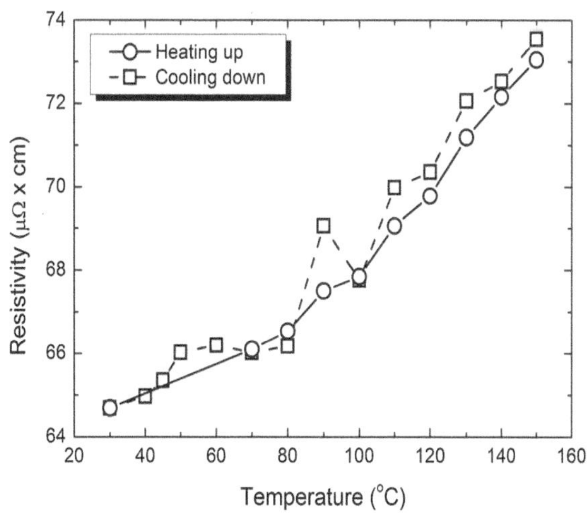

Fig. 2.16 Electrical resistivity variation of passivated 10 nm nickel film with temperature

Table 2.2 Performance data of the devices with ITO and Ni electrodes

	Work function (eV)	Current density (A m^{-2})	V_{TH}/V_{maxLce} (V)	Luminance (cd m^{-2})
Nickel 8 nm	5.1	5115	4.5/11	430
Nickel 10 nm	5.1	7880	4/8.5	330
ITO	5.0	7880	3.5/7.5	360

V_{TH} corresponds to the voltage at which 1 cd m^{-2} is obtained and V_{maxLce} corresponds to the voltage at which the maximum luminance is obtained

stability of the layer is confirmed by the absence of significant hysteresis, i.e., matching between "cooling-down" and "heating-up" curves. The calculated value of the temperature coefficient for the resistance (α) from Fig. 2.16 is 0.0106 °C^{-1}.

2.9 Use in OLED Device

Organic LEDs were the first devices to assess the potential of UTMFs. In collaboration with Stephanie Cheylan and Danny Krautz (ICFO), single component Ni UTMF was also used, instead of ITO, as a semitransparent hole injecting electrode for bottom polymer LEDs. Standard architecture of the diodes used for the study was: Anode/PEDOT:PSS/PFO/Al, where thermally evaporated Al layer (100 nm) was used as a cathode and the active layer Poly (9,9-dioctylfluorene) (PFO) was weighed and dissolved in chloroform and stirred overnight before spin coating to make the device.

The results comparing the performance of Ni based OLEDs with that to ITO are summarized in Table 2.2.

In terms of device efficiency, a slightly higher efficiency is obtained for ITO based device at low voltages. At higher voltages, Ni based devices show a similar or improved efficiency over the ITO based device. Confinement of light modes was also calculated for both ITO and Ni based devices using a program based on the matrix transfer method. Due to the fact that the Ni was extremely thin, no confined modes are present for it, while in the case of ITO, confinement of the order of 55–65 % for both TE and TM were obtained. This explains the fact that for Ni films, despite having lower transmittance compared to ITO, the device performance is as good as that of the ITO based device.

2.10 Conclusions

We have successfully deposited UTMFs on UV fused silica and silicon substrates with high uniformity and continuity. Surface analysis of the deposited samples show that their RMS roughness is always below 1 nm. Our results show that sufficiently thin Ni and Cr films are highly transparent in the shorter wavelengths (UV) region whereas ITO, due to its band gap, has strong absorption. In the IR region, metal layers show higher transparency while ITO has absorption due to resonance. In short, UTMFs show flatter optical transparency over the full wavelength range, from the UV to the IR (175 nm–25 μm), still maintaining high electrical conductivity. We have also solved potential issues due to oxidation of UTMFs by intentionally growing a few nanometers top oxide layer which in turn increases stability. The resulting stabilized films present higher optical transparency while preserving adequate electrical conductivity. The measured wide optical transmission and excellent electrical properties, combined with the proven stability after an *ad hoc* passivation treatment, make UTMFs serious competitors to ITO, in particular in the UV and IR ranges. We have also investigated the possibility of using single layer Ni UTMF as an alternative semitransparent electrode for OLED applications.

We have demonstrated that similar efficiencies are reached for devices with either ITO or Ni UTMF as bottom electrode. Such results are promising, also considering that the intrinsic properties of the Ni UTMFs are not yet optimized for device applications, as it is the case for ITO.

References

1. W.R. Grove, On the electro-chemical polarity of gases. Philos Trans R Soc Lond **142**, 87 (1852)
2. H. Bach, D. Krause, *Thin Film on Glass* (Springer, Berlin, 1997)
3. I.B. Freund, S. Suresh, *Thin Film Materials: Stress, Defect Formation and Surface Evolution* (Cambridge University Press, New York, 2003)
4. A. Goswami, *Thin Film Fundamentals* (New Age International, New Delhi, 1996)
5. M. Ohring, *The Materials Science of Thin Films* (Academic Press, San Diego, 1992)
6. M.P. Soriaga, *Thin Films*, vol. 25 (Springer, Berlin, 2002)
7. G. Hass, Structure and optics of evaporated metals films. Ann. Phys. **31**, 245 (1938)
8. R. Messier, Toward quantification of thin film morphology. J. Vac. Sci. Technol., A **4**, 3059 (1986)
9. J.A. Thornton, Influence of apparatus geometry and deposition conditions on the structure and topography of thick sputtered coatings. J. Vac. Sci. Technol. **11**, 666 (1974)
10. J.A. Thornton, Study of the microstructure of thick sputtered coatings. Thin Solid Films **40**, 335 (1977)
11. K. Wasa, S. Hayakawa, *Handbook of Sputter Deposition Technology: Principles, technology and Applications* (Noyes Publications, Park Ridge, 1992)
12. www.ajaint.com
13. F.C. Frank, J.H. van der Merwe, One dimensional dislocations. I. Static theory. Proc. R. S. A **198**, 205 (1949)
14. M. Volmer, A. Weber, Nucleus formation in supersaturated systems. Z. Phys. Chem. **119**, 6274 (1926)
15. J.N. Stranski, L. Krastanov, Zur Theorie der orientierten Ausscheidung von Ionenkristallen aufeinander. Ber. Akad. Wiss. Wien **146**, 797 (1938)
16. E. Bauer, Phenomenological theory of crystal deposition on surfaces. Z. Kristallogr. **110**, 372 (1958)

17. S. Giurgola, P. Vergani, V. Pruneri, Ultrathin metal films as an alternative to TCOs for optoelectronic applications. Nuovo Ciemento B **121**, 887 (2006)
18. G. Govindaraj, V. Devenathan, Quantum size effects in thin metal films. Phys. Rev. B **34**, 5904 (1986)
19. Y. Choi, S. Suresh, Size effects on the mechanical properties of thin polycrystalline metals films on substrates. Acta Mater. **50**, 1881 (2002)
20. J.J. Thomson, Proc. Camb Philos Soc **11**, 119 (1901)
21. K. Fuchs, The conductivity of thin metallic films according to the electron theory of metals. Proc. Camb. Philos. Soc. **34**, 100 (1938)
22. E.H. Sondheimer, The mean free paths of electrons in metals. Adv. Phys. **1**, 1 (1952)
23. Y. Namba, Resistivity and temperature coefficient of thin metal films with rough surface. Jpn. J. Appl. Phys. **9**, 1326 (1970)
24. A.F. Mayadas, M. Schatzkes, Electrical resistivity model for polycrystalline films: the case of arbitrary reflection at external surfaces. Phys. Rev. B **1**, 1382 (1970)
25. O.S. Heavens, *Thin Film Physics* (Methuen, New York, 1973)
26. E.D. Palik, *Handbook of Optical Constants of Solids* (Academic Press, New York, 1985
27. C.G. Granqvist, A. Hultaker, Transparent and conducting ITO films: new developments and applications. Thin Solid Films **411**, 1 (2002)

Chapter 3
UTMF-Based Transparent Electrode Incorporating a Metallic Grid

3.1 Motivation

Among all the alternatives to ITO, UTMFs can overcome the high cost of raw materials and can be grown using a simple processing [1–4]. The electrical resistivity (ρ) of UTMFs can be lower than that of TCOs. However, if one wants to achieve high optical transmission, the thickness of UTMFs should be limited to several nanometers, at the expense of sheet resistance (R_S). This implies that the R_S, which is given by ρ/t, of UTMFs is relatively high compared to TCOs [5]. For example the R_S of 2 and 10 nm thick Ni films are in the range of 1000 and 50 Ω/sq, respectively, while R_S of 100 nm ITO film is in the 15–20 Ω/sq range. If the thickness of Ni is further increased, lower R_S values can be achieved at the expense of the optical transmission. In fact 40 nm films can achieve R_S values of about 5 Ω/sq but it becomes opaque. The larger the R_S value of a TE the higher the Ohmic (resistive) loss in the devices using the TE, thus the larger the power (energy) loss, consequently the probability that the device fails due to thermal loading in the TE area increases [6, 7]. Ohmic losses become critical for high current devices (e.g. laser sources and solar cells) while they are less critical for low current devices (e.g. integrated modulators and displays, which rely on capacitive charging)

Patterning of metals has been recently proposed to achieve an efficient transparent electrodes [8–11]. The semi-transparent metal electrodes are in the form of a nanometre scale periodically perforated dense metal on the substrate. By changing the aperture ratio and the thickness of the metal the optical transparency and the R_S can be tuned. In fact it is shown that values comparable to ITO can be achieved at the same time for optical transparency and R_S. The linewidth (i.e. the width of the regions where the metal is present) has to be sub-wavelength, to provide sufficient transparency and to minimize scattering. In addition the period of the structure must be sub-micrometer to ensure the uniformity of the current on the substrate surface (e.g. uniform injection or extraction of currents into or from

D. S. Ghosh, *Ultrathin Metal Transparent Electrodes for the Optoelectronics Industry*, Springer Theses, DOI: 10.1007/978-3-319-00348-1_3,

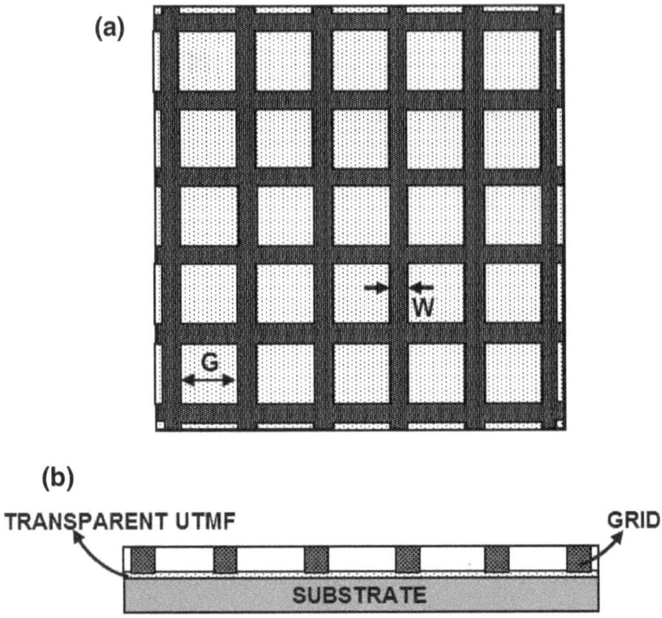

Fig. 3.1 **a** A drawing of the proposed grid pattern and **b** the cross-sectional view

an active layer). The latter drawback, i.e. the need of very small structural features with sub-micrometer periods, is due to the fact that the electrode is missing on the majority of the area and the Ohmic loss due to conduction has to be kept small. The structure being in the nanometer to micron range, the approach requires demanding and expensive conventional lithographic techniques for patterning. To make it less expensive nonconventional lithographic techniques, such as nano-imprinting, have been used. Such techniques have, however, several drawbacks, including a likelihood of missing metal lines and are relatively expensive than conventional techniques for patterning on a larger scale (micron to mm range), including UV lithography, screen printing and shadow masking during deposition.

In this chapter, a design is proposed and experimental results are reported in support of it which confirm that it is possible to reduce the R_S of UTMF based TEs of more than two orders of magnitude without significantly affecting their transparency. The proposed grid assisted UTMF based TE (G-UTMF) consists of a uniform UTMF with a top metal grid structure in contact with it (Fig. 3.1). However, the concept can be applied to either a different conductive material (semiconductor) or a different geometry (random mesh). By changing the line-width and the thickness of the grid, the optical transparency and the R_S can be tuned accordingly.

3.2 Design of the Grid

The sheet resistance of the proposed G-UTMF can be calculated considering a portion ABCD (Fig. 3.2) of the structure. The resistance of the portion ABCD is the sheet resistance of the whole structure G-UTMF. The portion ABCD can be further divided into five parts denoted by 1–5 as shown in Fig. 3.2

Let, ρ_G = resistivity of the thick film,
ρ_{UTMF} = resistivity of the UTMF,
t_G = thickness of the thick film,
t_{UTMF} = thickness of the UTMF,
The Sheet resistance of G-UTMF,

$$R_{S,TOT} = (R_4 + R_5 + R_2)\ ll\ R_1\ ll\ R_3 \tag{3.1}$$

where $R_4 = R_2$ and $R_1 = R_3$

Using the definition of resistivity and considering Ohm's law, $R_{S,TOT}$ can be calculated as follows

$$R_{S,TOT} = \xi \frac{\frac{\rho_G}{t_G}\frac{1}{f_F}\left(\frac{\rho_G}{t_G}\frac{W}{G} + \frac{\rho_{UTMF}}{t_{UTMF}}\right)}{\frac{\rho_G}{t_G}\frac{1}{f_F} + \frac{\rho_G}{t_G}\frac{W}{G} + \frac{\rho_{UTMF}}{t_{UTMF}}} \tag{3.2}$$

where ξ is a correction factor, which depends on deposition conditions and can be determined experimentally for a given deposition technique and specific process. G is the grid spacing and W is the linewidth. f_F is the filling factor defined by $W/(G + W)$. The filling factor thus quantifies the area covered by metal strips

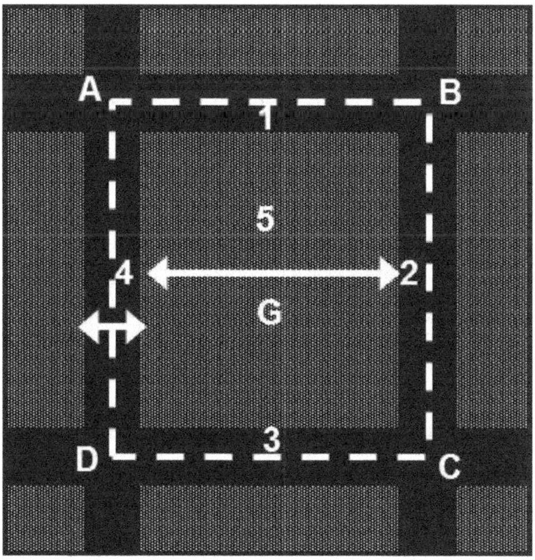

Fig. 3.2 A portion of the G-UTMF

compared to UTMF. In fact with good approximation the sheet resistance of G-UTMF on a scale larger than the period of the grid can be expressed by the following

$$R_{R,TOT} = \xi \frac{\rho_G}{t_G f_F} \tag{3.3}$$

Similarly, the optical transparency of the G-UTMF can be written down as follows

$$T_{TOT} = T_{UTMF} \times \left(\frac{G}{G+W} \right)^2 \tag{3.4}$$

Equation 3.4 can be rewritten in terms of filling factor as

$$T_{TOT} = T_{UTMF} \times (1 - f_F)^2 \tag{3.5}$$

where we have assumed that the thickness of the metal grid is large enough to make it opaque. From Eq. 3.5 one can gather that the transparency of the G-UTMF is independent of the grid metal thickness, only a function of the filling factor f_F. This eliminates the inverse trade-off relationship between optical transparency and electrical conductivity of metal based transparent electrodes.

One aspect of the structure relies on the use of a uniform, continuous UTMF with a larger thickness metal grid on top of it. Depending on the application such grid can be either on a nanometre to micron scale (e.g. in the case of LEDs where the scattering due to metal lines are to be avoided to preserve the output optical beam quality) or micron to hundreds of microns scale (e.g. in the case of solar cells where preserving the quality of the incoming optical beam is not essential). The period and linewidth of the grid depend also on the thickness of the underneath UTMF and the local currents involved. In general, the thinner the UTMF and the larger the current densities, the smaller the regions not covered by the grid.

In such a combined structure (G-UTMF), besides contributing to the collection and injection of electrical charges where the grid lines are absent, the underlying UTMF layer can also be used for other functionalities, for example, work function matching with active layers in photovoltaic and LED devices. Since the sheet resistance of the underlying UTMF becomes negligible on a scale larger than the grid period, the overall current distribution is mainly dominated by the metal grid-structure (Fig. 3.3).

Figure 3.4 shows the calculated dependence of $R_{S,TOT}$ on the linewidth for different grid spacing (t_G = 500 nm fixed). All of these calculations were carried out considering the same material (Ni) for both grid and UTMF. As the grid spacing widens, the sheet resistance slightly increases due to the decreased filling factor. Figure 3.5a shows the dependence of $R_{S,TOT}$ on f_F for different grid metal thicknesses (G = 500 μm fixed). With the increase in grid metal thickness, $R_{S,TOT}$ decreases significantly as the conductivity behavior relies on the thick metal grid-structure. The influence in f_F on the transparency is shown in Fig. 3.5b, exhibiting a linear behavior over the presented range. From this data it is evident that

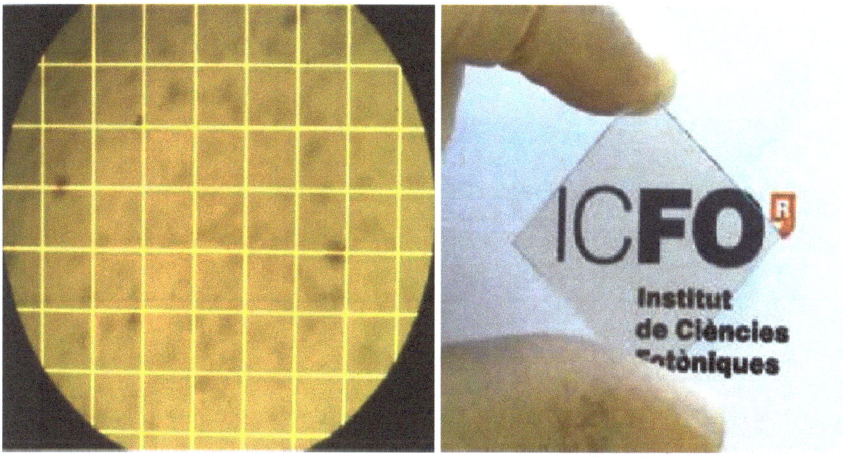

Fig. 3.3 Microscopic (*left*) and macroscopic (*right*) view of a G-UTMF

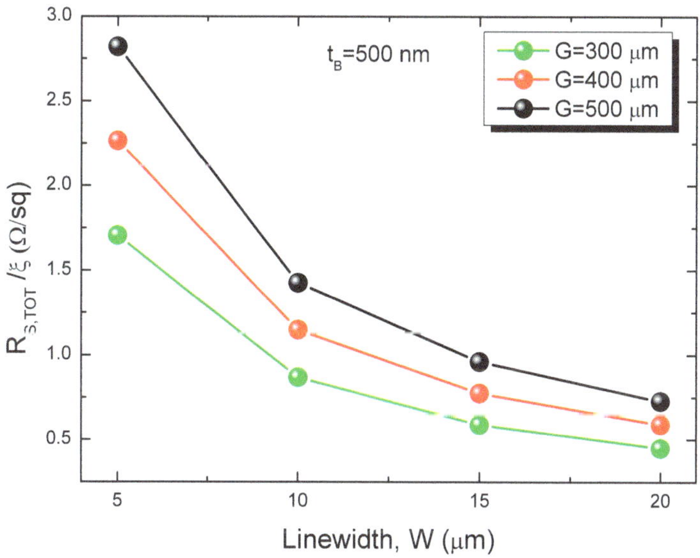

Fig. 3.4 Sheet resistance of G-UTMF as a function of linewidth for different grid spacing with fixed $t_G = 500$ nm

a trade-off exists between T_{TOT} and $R_{S,TOT}$, which corresponds to an optimum f_F. To define the trade-off we introduce the figure of merit, ϕ_{TE}, defined by Haacke [12].

$$\emptyset_{TE} = \frac{T_{TOT}^{10}}{R_{S,TOT}} \tag{3.6}$$

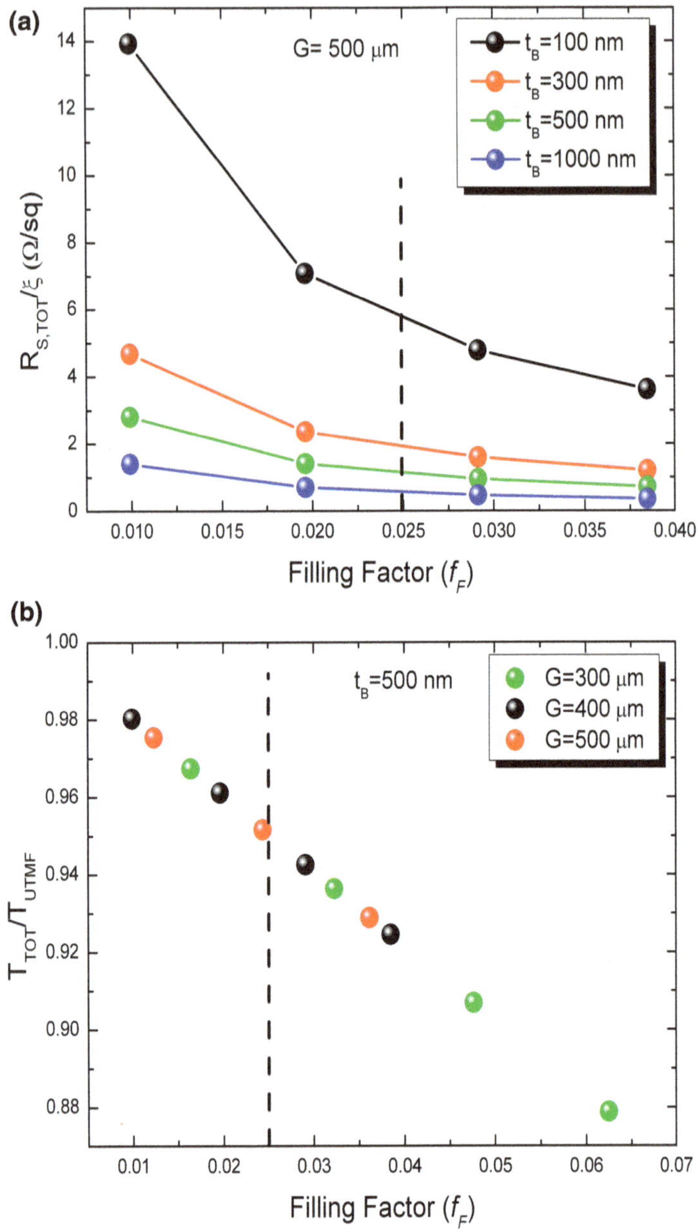

Fig. 3.5 Variation of sheet resistance **a** and transparency **b** of the G-UTMF with filling factor. The dotted line in the figures correspond to the calculated optimum filling factor

Using Eqs. 3.3 and 3.5 it can be rewritten as

$$\begin{aligned} \emptyset_{TE} &= \frac{t_G \times T_{UTMF}^{10} \times f_F \times (1 - f_F)^{20}}{\varepsilon \times \rho_G} \\ &= B \times f_F \times (1 - f_F)^{20} \end{aligned} \tag{3.7}$$

where B can be viewed as a constant for a given grid thickness, which equals to $\left[t_G \times T_{UTMF}^{10} / (\varepsilon \times \rho_G)\right]$. Considering the small value of f_F ($\ll 1$), Eq. 3.7 can be simplified by using Newton's binomial theorem as follows

$$\emptyset_{TE} \approx B\left(f_F - 20 f_F^2\right) \tag{3.8}$$

For this quadratic equation, the value of filling factor for which the ϕ_{TE} is maximized can be calculated as follows

$$\begin{aligned} \frac{d\emptyset_{TE}}{df_F} &= B(1 - 40 f_F) = 0 \\ f_F &= \frac{1}{40} = 0.025 \end{aligned} \tag{3.9}$$

From the definition of f_F, for the highest ϕ_{TE}, the ratio of spacing (G) to the linewidth (W) should equal to 39. Moreover, one can further tune the sheet resistance by proper choice of grid thickness, t_G and using different metallic materials.

3.3 Implementation

To realize the G-UTMF structure, a 2 nm Ni UTMF was deposited at room temperature on a double side polished UV fused silica substrate using 100 W DC power (Ajaint Orion 8 HV sputtering system) in a pure Ar atmosphere at a pressure of 2 mTorr. Prior to the deposition, the substrates were cleaned in standard way as mentioned in Chap. 2. A negative photoresist was then spin-coated on the film with 4000 rpm and dwell time of 30 s. This recipe produces a photoresist thin film of about 1.2–1.3 μm. In fact, the thickness of the photoresist determines the maximum thickness of the metal grid to be deposited later in the process. The sample was then baked at 120 °C for 2 min in order to evaporate the solvent from the photoresist. Thereafter, the sample was exposed to UV light for 5.1 s using Quintal Q4000 mask aligner with a hard mask of desired grid pattern placed on top of the sample. The sample was reversed baked for 1 min 30 s at 120 °C followed by flood UV exposure for 18 s. The exposed sample was then developed for about 40 s which removes the photoresist in the regions of the sample where the UV light is blocked by the hard mask, thus making openings in the sample to deposit the thick metal grid. The sample was again loaded in the sputtering chamber and a thick metal was deposited using the same aforementioned conditions. After

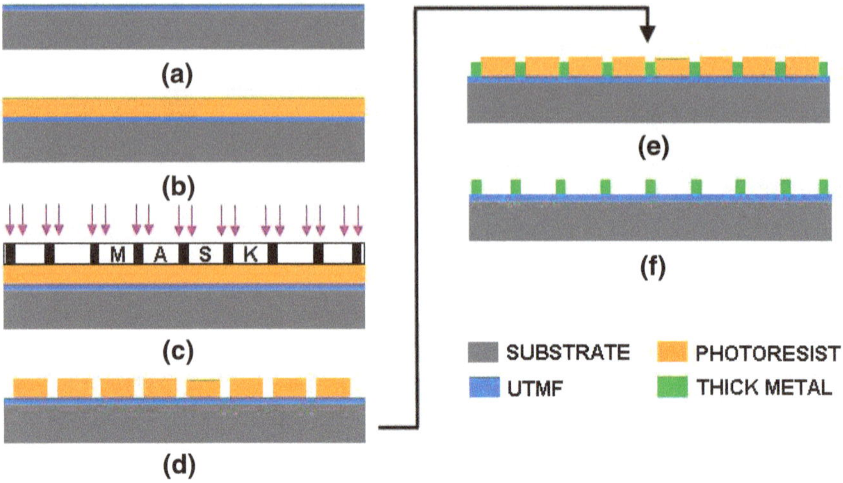

Fig. 3.6 Steps involved for the fabrication of G-UTMF

depositing the thick metal the sample was dipped in acetone and placed in an ultrasonic bath till the remaining photoresist was lifted off (Fig 3.6).

The electrical sheet resistance of the obtained G-UTMF and the initial UTMF was measured by Van der Pauw method and Four point Probe (Cascade Microtech 44/7S), respectively, as the latter method was found to be not always stable and reproducible for G-UTMF structures. For the transmission spectra, Perkin Elmer Lambda 950 spectrophotometer was used.

3.4 Experimental Results

To experimentally assess the proposed design, we fabricated four samples of different linewidth (5, 10, 15, and 20 μm), with fixed $G = 500$ μm and $t_G = 50$ nm. Both the UTMF and the grid were made of Ni. According to our deposition conditions, the value of ξ was determined as 7.2. Table 3.1 compares the calculated and experimental data, which are in good agreement for large *G/W* values. For small *G/W* values the approximation condition in Eq. 3.3 is no longer valid. However, the above-mentioned equations help to design and predict the performance of G-UTMF based transparent electrodes. Note that the grid thickness

Table 3.1 Comparison of experimental results with calculated theoretical values for different G/W ratios. Ni was used for both UTMF and grid material

G/W	25	33	50	100
Experimental data (R_S, T_{TOT})	(28.37, 76.5)	(51.4, 76.67)	(99.3, 76.83)	(197.16, 80.37)
Calculated data (R_S, T_{TOT})	(52.0, 74.25)	(67.7, 75.87)	(101.3, 77.2)	(197.76, 78.73)

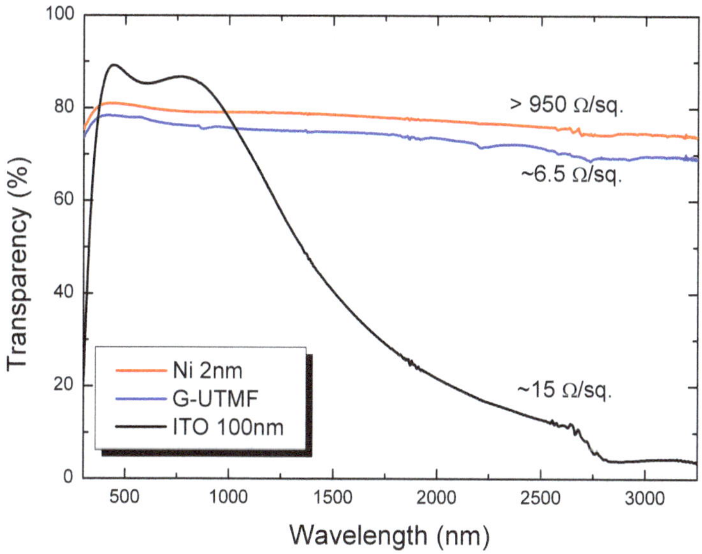

Fig. 3.7 Comparison of optical transparency for UTMF (2 nm Ni), G-UTMF (2 nm Ni + 100 nm Cu grid), and 100 nm ITO. The corresponding sheet resistances are also indicated. Note that the substrate contribution is taken into account as mentioned in Chap. 2

employed in Table 3.1 is fixed at 50 nm. One can increase the grid thickness and choose other metals such as copper, silver, or gold in order to reduce sheet resistance to below 10 Ω/sq without compromising the transparency.

To demonstrate the possibility, a G-UTMF sample using 100 nm Cu grid + 2 nm Ni UTMF was fabricated and a R_S of approximately 6.5 Ω/sq and optical transparency exceeding 75 % was achieved. Figure 3.7 shows the transparency dependence on wavelength of the Ni UTMF, Ni/Cu G-UTMF and, for comparison, of a standard ITO film. One can readily appreciate that the transparency of all the films is similar in the visible region while it is much higher for the metal based TEs in the UV and infrared, the latter aspect being crucial in several applications (e.g., UV and infrared detectors and photovoltaic cells for space).

3.5 Conclusions

In summary, by incorporating a thick metal grid on the top of an UTMF, an effective approach to reduce significantly (more than two orders of magnitude) the electrical sheet resistance of metal based TEs, with negligible loss in optical transparency is demonstrated. The theoretical calculation showed that the value of the filling factor and the ratio of grid spacing to linewidth should be around 0.025

and 39, respectively, to achieve the highest figure-of-merit. Further tailoring in optical and electrical properties of the proposed G-UTMF based TEs can be achieved by using different materials and grid geometries. It should be noted that the grid or mesh structure could be deposited using inexpensive techniques, such as screen printing and shadow masking during deposition. The process simplicity and the low cost of the proposed structures will have an impact in a wide range of applications, including photovoltaic cells, optical displays, detectors and electrochromic devices.

References

1. S. Giurgola, P. Vergani, V. Pruneri, Ultrathin metal films as an alternative to TCOs for optoelectronic applications, Nuovo Ciemento B 121, 887 (2006)
2. K.V. Rajani, S. Daniels, P.J. Mcnally, F.O. Lucas, M.M. Alam, Ultrathin chromium transparent metal contacts by pulsed dc magnetron sputtering. Phys. Status Solidi A **207**, 1586 (2010)
3. B. O'Connor, C. Haughn, K.H. An, K.P. Pipe, M. Shtein, Transparent and conductive electrodes based on unpatterned, thin metal films. Appl. Phys. Lett. **93**, 223304 (2008)
4. M. Avrekh, B.M. Thibadeau, O.R. Monteiro, I.G. Brown, Transparent, conducting, metallic thin films. Rev. Sci. Instrum. **11**, 4328 (1999)
5. S. Giurgola, A. Rodríguez, L. Martínez, P. Vergani, F. Lucchi, S. Benchabane, V. Pruneri, Ultra-thin nickel transparent electrodes. J. Mater. Sci. Mater. Electron. **20**, S181 (2009)
6. T. Minami, Present status of transparent conducting oxides thin-film development for ITO substitutes. Thin Solid Films **516**, 5822 (2008)
7. F. Martin, E. Navarrete, J. Morales, C. Roldan, J.R.R. Barrado, L. Sanchez, High-energy, efficient and transparent electrode for lithium batteries. J. Mater. Chem. **20**, 2847 (2010)
8. M.G. Kang, L.J. Guo, Nano imprinted semitransparent metal electrodes and their applications in organic light emitting diodes. Adv. Mater. **19**, 1391 (2007)
9. M.G. Kang, M.S. Kim, J. Kim, L.J. Guo, Organic solar cells using nano imprinted transparent metal electrodes. Adv. Mater. **20**, 4408 (2008)
10. M.G. Kang, L.J. Guo, Semitransparent Cu electrode on a flexible substrate and its application in organic light emitting diodes. J. Vacuum Sci. Technol. **B25**, 2637 (2007)
11. P.B. Catrysse, S. Fan, Nanopatterned metallic films for use as transparent conductive electrodes in optoelectronic devices. Nano Lett. **10**, 2944 (2010)
12. G. Haacke, New figure of merit for transparent conductors. J. Appl. Phys. **47**, 4086 (1976)

Chapter 4
Copper Bilayer Transparent Electrodes

4.1 Motivation

The shortcomings of ITO have led to the search for alternatives, such as single walled CNTs, graphene films, and UTMFs [1–4]. Copper in thin-film form is a very inexpensive material with excellent electrical and optical properties hence its wide use in microelectronics. However ultrathin (i.e. $\leq$10 nm) Cu films are prone to oxidation and corrosion, which alter significantly their electrical and optical properties. This chapter describes the development of bilayer transparent electrode (TE) structures consisting of an ultrathin and continuous Cu film covered by a protective ultrathin Ti or Ni film. The capping layer is chosen according to the functional requirements of specific devices. The top Ti or Ni capping layers are then typically in situ treated with O_2 plasma which notably increase their transmittance. Both the Ni and Ti enhance the stability of the Cu based TEs and, in addition, Ni also increases the work function. The measured optical performance of the fabricated films is in good agreement with theoretical estimation based on multiple reflection and refraction (MFMR) model [5, 6]. The obtained Cu-Ti/Ni bilayered structure meets the requirements of the majority of optoelectronics applications, such as organic LEDs and photovoltaic solar cells, liquid crystal displays, presenting large optical transparency, low electrical R_S, and stability against temperature and oxidation. Transparency as high as 86 % at 630 nm, R_S as low as 16 Ω/sq have been achieved in bilayer Cu based films, which also exhibit significantly higher stability than single layer Cu films against temperature in ambient atmosphere.

4.2 Implementation

Polycrystalline UTMFs were deposited by DC magnetron sputtering (ATC Orion 8 HV) on double-side optically polished UV grade silica substrates. Prior to the deposition, the substrates were cleaned in standard way as mentioned in Chap. 2,

D. S. Ghosh, *Ultrathin Metal Transparent Electrodes for the Optoelectronics Industry*, Springer Theses, DOI: 10.1007/978-3-319-00348-1_4,

Sect. 2.2. The deposition rate was determined to be 1.5 Å/s for Cu, 0.083 Å/s for Ti and 0.573 Å/s for Ni, respectively. We prepared the following four sets of samples: Cu of different thickness, Cu of different thickness with a constant 1 nm Ni capping layer, Cu of different thickness with a constant 1 nm Ti capping layer, and Cu of different thickness with a constant 5 nm Ti capping layer, named Cu(δ), Cu(δ) + Ni(1), Cu(δ) + Ti(1), and Cu(δ) + Ti(5), respectively, where δ is the thickness in nanometers. The Cu(δ) + Ti(5) samples have been further treated in situ after the deposition. The treatment consisted of exposing the samples to O_2 plasma for 15 min under a working pressure of 8 mT and 40 W RF power. The treated samples have been named Cu(δ) + Ti(5)_TR. A Perkin Elmer lambda 950 spectrometer was used for optical transmission measurements while Cascade Microtech 44/7 S 2749 four-point probe system with a Keithley 2001 multimeter for electrical sheet resistance (R_S) measurements. The morphology of the fabricated films was characterized by AFM with the digital instrument Veeco Digital Instrument Dimension 3100 AFM and associated software.

4.3 Experimental Results and Analysis

Figure 4.1 compares the performance of Cu based transparent UTMFs alongside SWCNT and graphene film, all of them considered alternative transparent electrodes (TEs) of TCOs. It can be seen that the Ti capping layer increases the transparency without any significant change in RS while the Ni capping decreases the transparency by ~7–8 % as Ni has higher extinction coefficient compared to Ti. For UTMFs, the theoretical models for resistivity do not work very well, because besides geometrical limitation (size-effect) that enhances surface scattering, the resistivity also depends on volume sources of scattering, such as

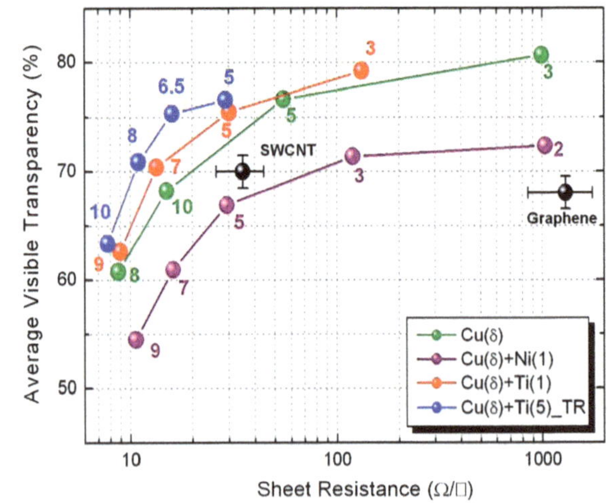

Fig. 4.1 Visible optical transparency as a function of R_S for different sample sets, Cu(δ), Cu (δ) + Ni1, Cu (δ) + Ti1, and Cu(δ) + Ti5_TR. The value of Cu thickness (δ) identifies each point in the curve. The corresponding values of SWCNT and graphene are also shown

grain boundaries, voids, and discontinuities, which are very sensitive to deposition conditions as explained in Chap. 2, Sect. 2.6.1. Therefore, in our case, the deposition conditions were optimized through subsequent iterations, including the process for O_2 plasma treatment, so that the films have the lowest R_S. It is observed that the conductivity behavior of the four sample sets is mainly dominated by the underlying ultrathin Cu layer when it becomes continuous (Cu thickness >5 nm). Assuming that the total resistances of the bilayer coupled in parallel, it is possible to calculate the specific resistance of the bilayer structure. This assumption is only justified if film boundary effects are negligible [7, 8]:

$$R_S = R_{Cu} \times \frac{R_{Cap}}{\left(R_{Cu} + R_{Cap}\right)} = \frac{R_{Cu}}{1 + \eta} \tag{4.1}$$

where $\eta = R_{\mathrm{Cu}}/R_{\mathrm{Cap}}$, and R_{S}, R_{Cu}, R_{Cap} are the sheet resistance of the bilayer structure, the conductive Cu layer, and the capping layer, respectively. The O_2 plasma treatment on Cu(δ) + Ti(5) oxidizes only the top few nanometers of Ti without altering the interface between Cu and Ti. The formation of titanium oxide not only protects the underlying Cu layer from oxidation but also increases the optical transmission of the bilayer structure. We have also tried depositing TiO_2 directly from a target and found that directly depositing an oxide layer on the top of Cu UTMF can inevitably lead to partial oxidation of the Cu layer, with a consequent increase in the R_S. Moreover, according to Eq. 4.1, the presence of the titanium oxides capping layer should have negligible influence on the overall R_S given the fact that $R_{Cap} \gg R_{Cu}$. In other cases, Ni is chosen as capping metal layer since it has high resistant to oxidation as well as high work function (5.15 eV). For some specific applications such as OSCs and OLEDs, the high work function of the electrode plays an essential part in final device performance when the TE is used as an anode.

To compare the performance of different sample sets, we calculated the figure-of-merit, Φ_{TC}, which takes into consideration the transparency and the R_S of the electrode, according to the definition provided by Haacke [9], $\Phi_{TC} = T_{\mathrm{lum}}^{10}/R_{\mathrm{S}}$, where T_{lum} is the photopic transparency, i.e., the transparency normalized by photopic luminous efficiency function as follows:

$$T_{lum} = \int_{375}^{780} f(\lambda)T(\lambda)d\lambda \Big/ \int_{375}^{780} f(\lambda)d\lambda \tag{4.2}$$

f(λ) being the relative sensitivity of the human eye in the photopic state and *T(λ)* being the transparency spectrum of the films.

Figure 4.2a shows Φ_{TC} for the different sample sets. Among the four sets of samples, the Cu(δ) + Ti(5)_*TR* samples present a peak value of Φ_{TC} equal to $4.1 \times 10^{-3}\ \Omega^{-1}$. Evaluating from the data presented in Fig. 4.1, Φ_{TC} of the proposed ultrathin Cu based TEs is significantly higher than that of SWCNT and graphene films, currently considered as potential replacements of widely used TCOs. For all the sample sets the optimum Φ_{TC} corresponds to a Cu thickness between 5.5 and 6.5 nm which indicates that Cu films become continuous in this thickness range. The percolation threshold, i.e., the thickness corresponding to which the film changes from an island distribution to a continuous layer, is also estimated by plotting $R_S\delta_{tot}^3$ versus δ_{tot} (where δ_{tot} stands for the total film thickness) for the different sample sets (Fig. 4.2b) [10]. The percolation threshold for all the sets is found to be between 5.5 and 6.5 nm, which confirms the film continuity thickness resulting from the Φ_{TC} curves. In addition, the peak-to-valley value measured by AFM for the samples with Cu thickness less than 5 nm is higher than the film thickness, which implies that the films are discontinuous over the microscopic dimension (Fig. 4.3).

In order to better understand the enhanced figure-of-merit in the Cu + Ti(5)_TR structure, we calculated the reflection and transmission of our films based on the MFMR model, in which multiple reflection and refraction are taken into account. According to the MFMR model, the total reflectance (*R*) and transmittance (*T*) for a N-layer system can be written as

$$R_{N-j+1} = \frac{r_j + R_{N-j}e^{2i\Delta_j}}{1 + r_jR_{N-j}e^{2i\Delta_j}} \tag{4.3}$$

$$T_{N-j+1} = \frac{t_jT_{N-j}e^{i\Delta_j}}{1 + r_jR_{N-j}e^{2i\Delta_j}} \tag{4.4}$$

where r_j and t_j are the Fresnel coefficients for $j = 1,\ldots,N$ and $\Delta j = 2\pi\tilde{n}_jd_j/\lambda$ is the phase difference between boundaries of the *j*th layer, $\tilde{n}_j$ being the complex refractive index ($\tilde{n}_j = n_j + ik_j$) and d_j being the thickness of the *j*th layer. In our case, we considered a two-layer system and carried out calculations based on normal light incidence, smooth interface, and uniform refractive index of each layer. Figure 4.4 compares the calculated and experimental data for Cu(6.5) and Cu(6.5) + Ti(5)_TR samples, respectively. Compared to pure Cu(6.5), the Cu(6.5) + Ti5_TR shows suppressed reflection as evidenced from both

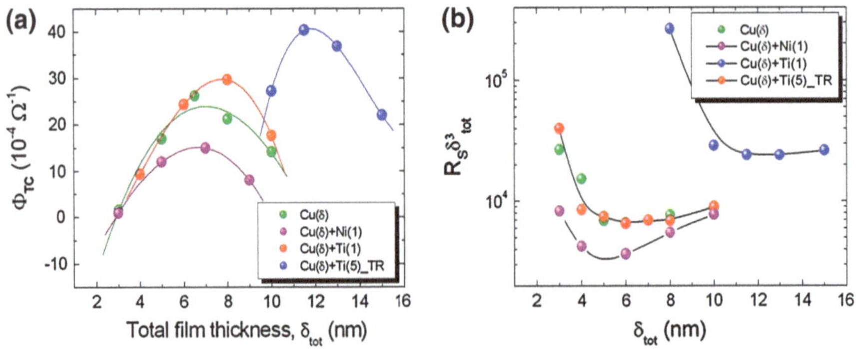

Fig. 4.2 **a** Figure-of-merit (Φ_{TC}) for different sample sets, Cu(δ), Cu (δ) + Ni1, Cu (δ) + Ti1, and Cu(δ) + Ti5_TR. The total thickness (δ_{tot}) is the sum of deposited Cu and Ni/Ti thicknesses, **b** determination of percolation threshold. The value of Cu thickness (δ) identifies each point in the curve

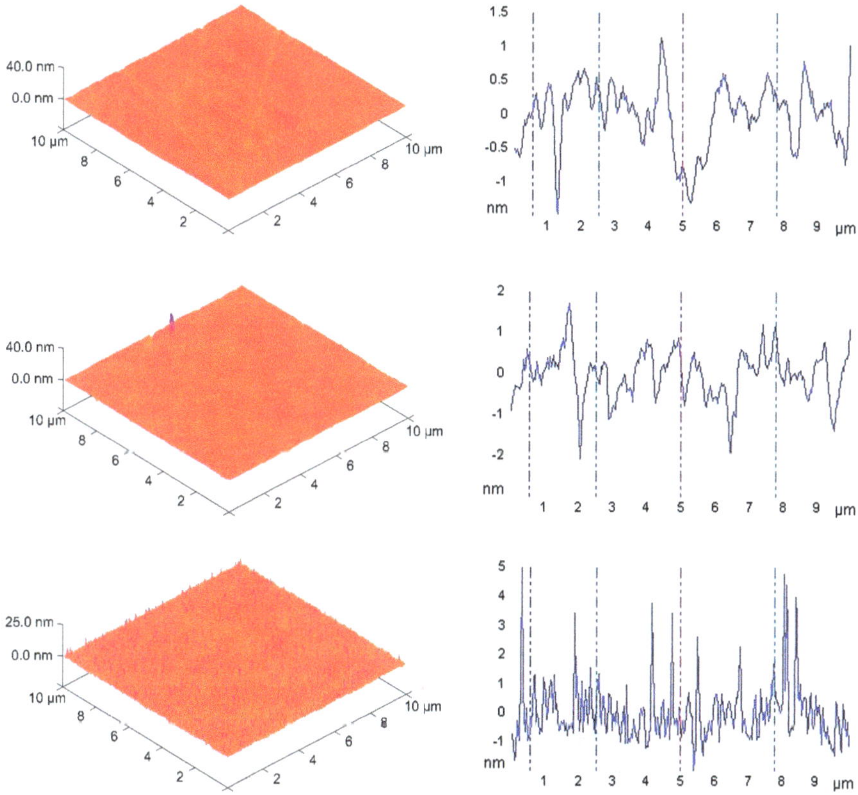

Fig. 4.3 Three dimensional AFM and thickness profile images of Cu(6.5), Cu(6.5) + Ni(1) and Cu(6.5) + Ti(5)_TR respectively from top to bottom. The RMS roughness of the films is determined to be 0.7 nm, 0.9 nm and 0.82 nm for Cu(6.5), Cu(6.5) + Ni(1) and Cu(6.5) + Ti(5)_TR respectively

experimental data and theoretical prediction (Fig. 4.4a), thus leading to larger transparency. Figure 4.4b depict the experimental and calculated transparency as a function of wavelength for Cu(6.5) and Cu(6.5) + Ti(5)_TR sample, respectively. There is a good agreement between the calculated and experimental values, which suggest that the proposed model can be used as a potential design tool for such bilayer metallic TEs. It is obvious from both Figs. 4.1 and 4.4 that Ti capping layer in situ treated by O_2 plasma can apparently improve the optical properties of Cu based UTMFs due to contribution of multiple reflection and refraction effects in Cu-Ti system.

To evaluate the stability of the proposed Cu based TEs, we fabricated three sample sets, Cu(8), Cu(8) + Ni(1), Cu(8) + Ti(1), and Cu(8) + Ti(5)_TR and kept them in ambient air for a dwell time of 45 min at several fixed temperatures (T) and measured the $R_S(T)$ after each thermal cycle at room temperature. The stability of the Cu based TEs can be evaluated by the relative slope variation

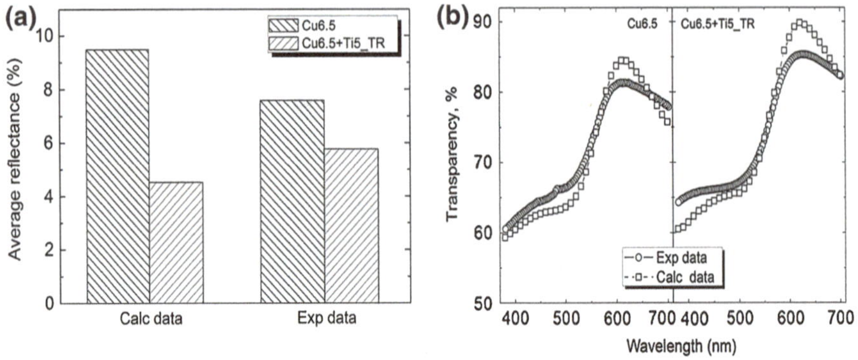

Fig. 4.4 **a** Average visible reflection from the experimental and calculated data for Cu (6.5) and Cu(6.5) + Ti(5)_TR samples, **b** experimental and calculated transparency for Cu (6.5) and Cu(6.5) + Ti(5)_TR samples respectively

$$tan\theta = \frac{|R_S(T - \Delta T) - R_S(T)|}{\Delta T} \tag{4.5}$$

By polar-plotting θ versus T, as it is shown in Fig. 4.5, the stability can be readily identified by the relative position in the figure. The ideal stability corresponds to the horizontal axis $\theta = 0$, which implies a R_S independent of temperature. Compared to Cu(8) and Cu(8) + Ti(1) samples which are subjected to an evident increase in R_S after thermal treatment, the Cu(8) + Ti(5)_TR sample is found to be more stable, while Cu(8) + Ni(1) is little influenced by temperature, exhibiting excellent thermal stability up to >180 °C.

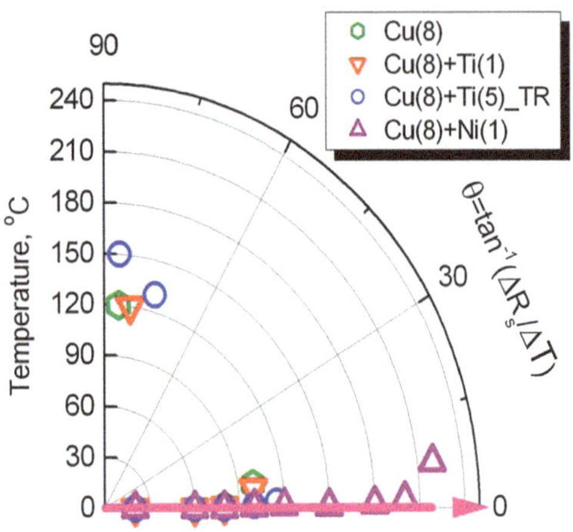

Fig. 4.5 Polar figure for the stability of the proposed Cu based TEs. The horizontal axis ($\theta = 0$) corresponds to ideal stability

4.4 Performance in OLED and OPV Devices

Out of the four sets of sample investigated, Cu(δ) + Ni(1) was then examined for possible use in OPV and OLEDs as it has high work function required to be used as an anode. In collaboration with Stephanie Cheylan and Danny Krautz (ICFO), as deposited Cu(δ) + Ni(1) films were used, instead of ITO, as transparent hole-injecting electrodes in bottom emitting polymer LEDs. Standard architecture of the diodes used was: Anode/PEDOT:PSS/SY/CaAg. Poly(phenylenevinylene) copolymer (SY), commercially available from Merck, is dissolved in toluene, stirred overnight and filtered before being spin-coated onto the PEDOT-PSS layer to a thickness of around 80 nm. Ca and Ag are sequentially thermally evaporated ($<5 \times 10^{-6}$ mTorr) onto the emitting layer using shadow mask. As ozone treatment of anode is essential pre-requisite before PEDOT:PSS spin coating, its effect on Cu(δ) + Ni(1) is also investigated during device fabrication. It is found that the 15 min of ozone treatment has very negligible effect on its properties while pure Cu degrades both its electrical and optical properties confirming the importance of Ni capping layer (Table 4.1).

For comparison, commercially available ITO (100 nm and 17–21 Ω/sq) is used to realize the same device architecture. The Cu(7) + Ni(1) anode based device shows similar performance to the ITO-based one. The excellent device performance based on Cu(7) + Ni(1) anode is related to a significant improvement in the injection of charges due to an improved work function matching at the anode-organic interface and low R_S.

In addition, in collaboration with the Organic Nanostructured Photovoltaics group at ICFO, the developed TE has also been used in OSCs. The OSC had the following configuration: Anode/NiO_X/P3HT:PCBM/$LiCoO_2$/Al. Standard ITO devices were also made for comparison (Table 4.2).

In all devices, the values of open circuit voltage (Voc) and fill factor (FF) were practically the same. The differences between the devices were related to changes in short-circuit current density (J_{SC}), which affects directly the efficiency. However, the OSC device incorporating UTMF based electrodes could achieve an efficiency of about 77 % of that using an ITO. The high efficiency of the Cu(7) + Ni(1) was partly due to the high metal reflectance which in turn could give rise to an enhanced photon absorption, due to the formation of an optical microcavity between Cu-Ni anode and the Al cathode.

Table 4.1 Performance data of the devices with ITO and Cu based electrodes

	Work function (eV)	Current density (A/m^2)	V_{TH}/VmaxLce (V)	Luminance (cd/m^2)
Cu	4.65	5860	2.6/8.1	10020
Cu(7) + Ni(1)	5.1	8830	2.2/7.1	12500
ITO	5.0	8830	2.2/6.2	12500

V_{TH} corresponds to the voltage at which 1 cd/m^2 is obtained and V_{maxLce} corresponds to the voltage at which the maximum luminance is obtained

Table 4.2 List of electrical characterization data for the OSC device

	J_{SC} (mA/cm2)	V_{OC} (mV)	FF (%)	PCE (%)
ITO	−10.62	520	59	3.31
Cu(7) + Ni(1)	−8.26	505	59	2.51

4.5 Conclusions

In conclusion, an inexpensive, easy-to fabricate, ultrathin Cu based TEs stabilized using a capping Ti or Ni layer has been developed. An average visible transparency as high as 75 % (peak transparency of 86 %) and stability against temperature and oxidation up to 180 °C has been achieved. The capping layer thus acted as a functional layer and can be chosen depending on the application specific device requirements. The electrical conductivity of the electrodes is dominated by the underlying continuous Cu layer and sheet resistance as low as 16 Ω/sq were reached. The fig-of-merit of the proposed ultrathin Cu based TEs was significantly higher than that of SWCNT and graphene films, currently considered potential replacements of widely used TCOs. The OPV device fabricated on Cu(7) + Ni(1) as anode reaches 77 % of that using ITO showing its their effectiveness for organic electronics. The bilayer metallic concept presented in this Chapter can also be extended to other metals, for example Ag and Pt instead of Cu and Ti/Ni, respectively.

References

1. Z. Wu, Z. Chen, X. Du, J.M. Logan, J. Sippel, M. Nikolou, K. Kamaras, J.R. Reynolds, D.B, Tanner, A.G. Rinzler, Transparent, conductive carbon nanotube films. Science **305**, 1273 (2004)
2. X. Wang, L. Zhi, K. Mullen, Transparent, conductive graphene electrodes for dye-sensitized solar cells. Nano Lett. **8**, 323 (2008)
3. S. De, T.M. Higgins, P.E. Lyons, E.M. Doherty, P.N. Nirmalraj, W.J. Blau, J.J. Boland, J.N. Coleman, Silver nanowire networks as flexible, transparent, conducting films: extremely high DC to optical conductivity ratios. ACS Nano **3**, 1767 (2009)
4. S. Giurgola, P. Vergani, V. Pruneri, Ultrathin metal films as an alternative to TCOs for optoelectronic applications. Nuovo Ciemento **B 121**, 887 (2006)
5. O.S. Heavens, Optical properties of thin films. Rep. Prog. Phys. **23**, 1 (1960)
6. H. Hirayama, K. Kaneda, H. Yamashita, Y. Monden, Visualization of optical phenomena caused by multilayer films based on wave optics. Vis. Comput. **17**, 106 (2001)
7. J. Lewis, S. Grego, Highly flexible transparent electrodes for organic light-emitting diode-based displays. Appl. Phys. Lett. **85**, 3450 (2004)
8. A. Kloppel, A. Scharmann, Dependence of the electrical and optical behaviour of ITO–silver–ITO multilayers on the silver properties. Thin Solid Films **365**, 139 (2000)
9. G. Haacke, New figure of merit for transparent conductors. J. Appl. Phys. **47**, 4086 (1976)
10. A.I. Maaroof, B.L. Evans, Onset of electrical conduction in Pt and Ni films. J. Appl. Phys. **76**, 1047 (1994)

Chapter 5
Ultrathin Alloy Films as TEs

5.1 Motivation

UTMFs have been studied in organic optoelectronics as stand-alone transparent electrodes (TE) [1, 2] or in combination with conducting oxides [3, 4], as it has also been described in the previous chapters. UTMFs have cost and processability advantage over the TCOs, such as the widely used ITO. Although UTMFs has wider transparency spectrum from UV to mid-IR (Sect. 2.7), the spectrum within the visible range is non-uniform due to the plasma formation at a characteristic frequency $\left(\omega_p = (4\pi ne^2/m)^{1/2}\right)$ as well as intraband electronic transitions. The peak transmittance of UTMFs, which also provide sufficiently large electrical conduction, is usually as high as 80 % at the plasma frequency and it rolls off the sides of this value. This translates into both a lower average optical transmission as well as a non-uniform wavelength response. One solution to counteract the transmission decrease associated to reflection above the plasma frequency is to use antireflection thin film oxides, such as ZnO, ITO, and TiO_2. The resulting multilayer TE structures can present high optical transparency and low sheet electrical resistance [5, 6]. Their complexity has the drawback of increasing processing time and material cost while the brittleness of the oxides limits the mechanical flexibility. Another solution is to combine UTMFs which present transmission peaks at different wavelengths in the visible region. In that way the advantages of UTMFs, such as a good trade-off between transparency and sheet resistance (R_S), simple processing, low cost and mechanical flexibility, can be maintained while achieving a flat wavelength response in the visible region. For example bilayer metal electrodes for top emission organic light-emitting devices, which consists of thin layers of Ca and Ag metals of different thicknesses, have been investigated [7].

In this chapter, a novel approach, consisting in co-sputtering Ag and Cu. UTMFs of Ag and Cu show peak transmittance at around 335–630 nm, respectively is described. The forming Ag/Cu alloy presents a transmission spectrum flatter with respect to the individual Cu and Ag layers. In addition, since both Ag and Cu are UTMFs that exhibit very high electrical properties their mixture shows low R_S. If

D. S. Ghosh, *Ultrathin Metal Transparent Electrodes for the Optoelectronics Industry*, Springer Theses, DOI: 10.1007/978-3-319-00348-1_5,

need be in the device where they are used, the alloy films can be further coated with, for example a Ni layer, to enhance both their stability and work function as demonstrated in Chap. 4. The resulting alloy films are exceptionally smooth, truly flexible, and remarkably resistant at elevated temperatures and with high work function. To prove their device potentials, the developed UTMF alloys were used as a semitransparent anode in OPVs, showing equal PCE of similar devices incorporating ITO electrodes.

5.2 Optical Spectra

To predict and assess the potential of alloy films as semitransparent electrode for display, photovoltaic and lighting applications, we analyzed their optical behavior by solving Fresnel's equations, i.e. the matrix formalism described in Sect. 4.3.

The *n*, *k* parameters used for transmission calculation of alloy films were average weighted values of the constituent metal films at any given wavelength (Eqs. 5.1 and 5.2).

$$n_{Alloy,\lambda} = \left(n_{1,\lambda} + n_{2,\lambda}\right)/2 \tag{5.1}$$

$$k_{Alloy,\lambda} = \left(k_{1,\lambda} + k_{2,\lambda}\right)/2 \tag{5.2}$$

where (n_1, k_1) and (n_2, k_2) are the optical constants for the two constituent metal. Fig. 5.1 shows the theoretical prediction of optical transparency for different alloy

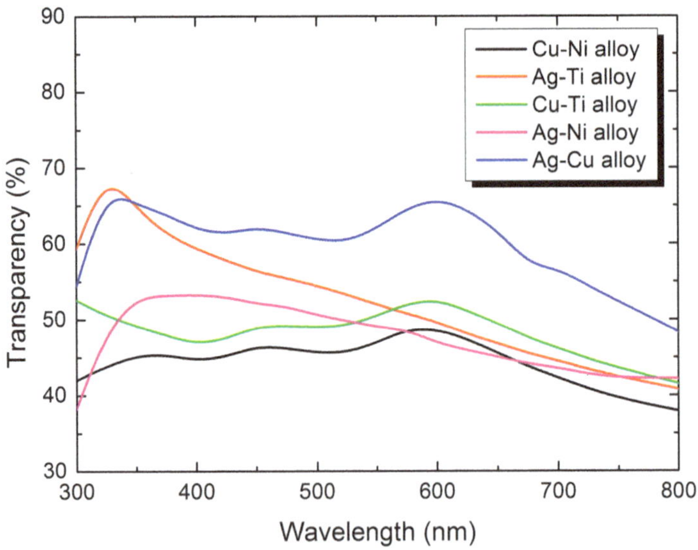

Fig. 5.1 Optical transmission in the visible spectrum for different alloy films with 50–50 % composition for the two constituent metals

films of 50–50 % composition in the visible range. For all the curves, 10 nm thick films were considered. As it can be seen in the figure, Ag–Cu alloy films have the highest and flattest optical transmission among all alloy films and therefore we have chosen them for this study.

We have also calculated the optical response of the Ag–Cu alloy films for different percentage compositions. As it can be seen in Fig. 5.2, the amount of Ag content dominates the optical behavior of the alloy film. For Ag composition $\geq$50 %, the alloy films begin to show higher average transmission in the visible region. The 50–50 % composition shows the flattest optical response

5.3 Implementation

The alloy films were prepared onto 1 mm-thick UV grade optically polished silica substrates by means of DC magnetron sputtering (AJA Orion 8 DC) in pure argon atmosphere. Prior to the deposition, the substrates were cleaned in a standard way, as it is described in Sect. 2.2. To prepare alloy films by co-sputtering, deposition rate of pure Cu and Ag targets were measured and the process was set up to achieve the same deposition rate for both metals. The thickness of the resulting alloy films is given by the sum of the individual metal thickness as the metals were sputtered on their own. Single component Cu and Ag films were also prepared under the same conditions to compare with the alloy films. It has to be said that

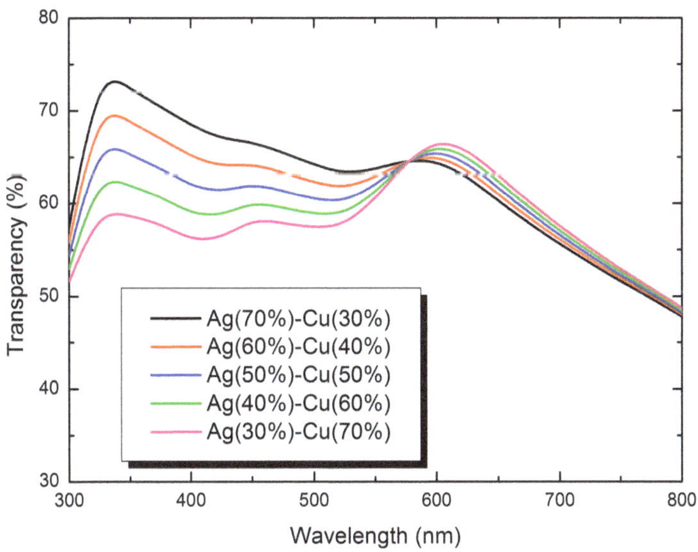

Fig. 5.2 Optical transmission for different percentage composition of Ag–Cu alloy films in the visible spectrum

another way of preparing alloy films is to sputter directly from an alloy target of a given composition, e.g. 50 % Cu and 50 % Ag. However this approach has the disadvantage of target degradation after a few depositions due to different deposition (target consumption) rates of the constituent metals. A Perkin Elmer lambda 950 spectrophotometer was used for optical transmission measurements while a Cascade Microtech 44/7 S 2749 four-point probe system with a Keithley 2001 multimeter for electrical sheet resistance measurements. The surface morphology and composition of the fabricated films were characterized with a digital instrument D3100- AFM and SEM, respectively.

5.4 Optical, Electrical and Mechanical Performance

As it can be seen in Fig. 5.3, there is good agreement between experimental and theoretical curves for the alloy film, despite the fact that we have considered bulk values for n, k of Cu and Ag, which can be different from the actual values when the thickness of the film becomes smaller than 10 nm [8]. As it was expected, the alloy film shows a quite flat behavior compared to single component Cu and Ag films, with the peaks due to Ag and Cu strongly reduced. As it will be shown in the next section this makes the alloy films particularly suitable for application-specific devices, such as OSCs (Fig. 5.4).

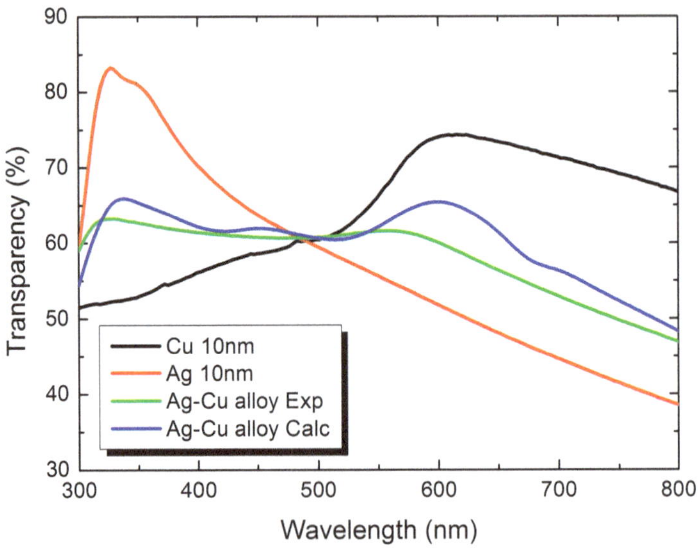

Fig. 5.3 Experimental comparison of optical spectra between Ag/Cu (50–50 %) alloy and single component Ag and Cu metal films. The figure also shows the theoretical optical prediction for the alloy film

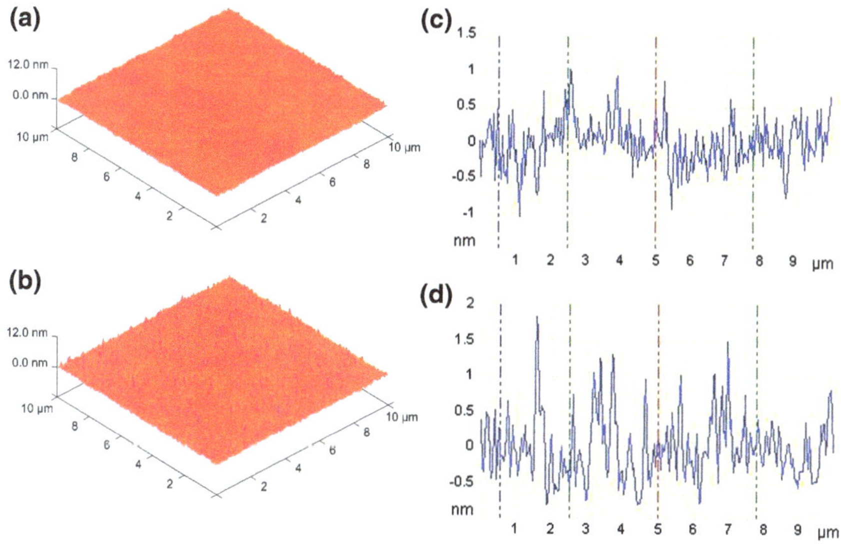

Fig. 5.4 Three dimensional AFM and thickness profile images of **a** Cu-Ag alloy and **b** Ni capped alloy film. The RMS roughness of the alloy film is 0.523 nm which is similar to single component Cu and Ag films, while the RMS roughness increases to 0.9 nm after coating with 1 nm Ni capping layer

Generally, alloy films are usually harder and less conductive than their constituent metals. Metals when deposited as single materials show uniform density while alloys can vary their density across the substrate, thus leading to enhanced scattering of conduction electrons, in turn larger electrical resistivity. This is confirmed by the fact that the sheet resistance (R_S) of the deposited alloy films is slightly larger than those of single component metal films. The R_S of 10 nm alloy film is about 15 Ω/sq while similar thicknesses of Cu and Ag films provide 12 and 8 Ω/sq, respectively.

Energy dispersive X-ray spectroscopy (EDX) was also carried out to determine the actual chemical composition of the deposited alloy films (Fig. 5.5). The films showed nearly equal presence of Cu and Ag. The AFM analysis of the deposited alloy films and Ni 1 nm capped films reveals RMS values for the roughness lower than 1 nm, which demonstrate an exceptionally smooth surface morphology.

The stability of the alloy films was assessed in harsh conditions, at high temperature and in the presence of oxygen atmosphere. The pure alloy films starts degrading in performance at 90 °C in ambient conditions while the Ni 1 nm capped alloy film maintains its electrical and optical properties up to elevated temperatures of about 180 °C, while still preserving its RMS roughness below 1 nm. In fact, the Ni capped alloy films behave similar to the Cu(δ) + Ni(1) films, already described in Chapter 4. Besides improving the durability of the underlying alloy, the 1 nm Ni capping layer increase the work function, being this feature crucial in the case of OSC devices. The developed alloy electrode also exhibited

Fig. 5.5 EDX analysis of the Ni 1 nm capped alloy films

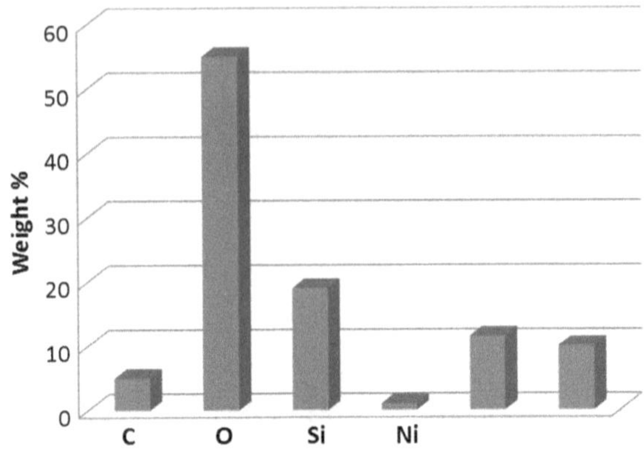

excellent mechanical flexibility, as it was evidenced by the observed crack-free behavior of deposited alloy on PET substrates, even after 10^4 bending cycles with a radius of curvature (r_C) continuously going from a minimum of 4 to a maximum of 12 mm. During the mechanical flexibility tests the electrical and optical properties were also maintained, while similar tests on commercially available ITO on PET substrate shows an increase in R_S of 550 % just after 10 cycles. This value increases to 1580 % after 20 cycles. In addition cracks become evident in the ITO film on the PET substrates after a few cycles, in agreement with the observed sheet resistance increase (Fig 5.6).

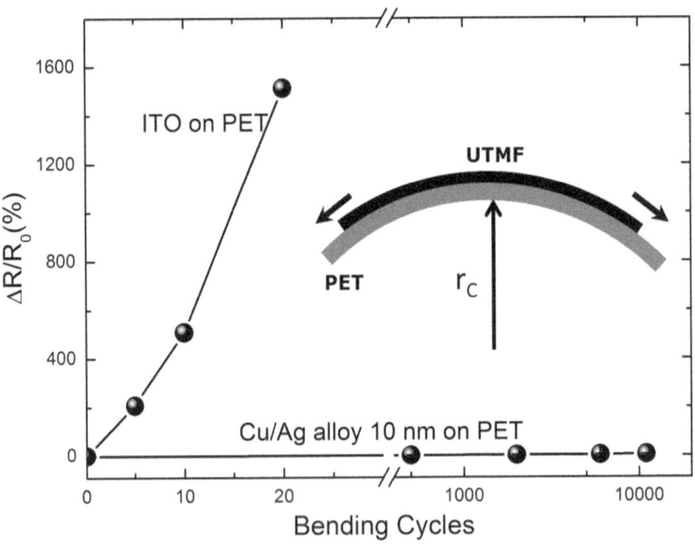

Fig. 5.6 Sheet resistance change during bending as a function of number of bending cycles for Cu-Ag alloy and ITO

5.5 Performance in an OPV Device

In collaboration with Jaehyung Hwang and Ingmar Bruder from BASF, the developed alloy TE has also been used in OSCs. The OSC had the following configuration: Anode/MoO_3(5 nm)/ID345:C60(1:1,35 nm)/C60(25 nm)/Bphen(5 nm)/Ag(100 nm). Devices incorporating standard ITO and single component Cu and Ag were also made for comparison.

The corresponding electrical characterization data the OSC devices integrating alloy electrode is summarized in Table 5.1. In Chapter 4, we have demonstrated working OSCs reaching 77 % of PCE of ITO based device utilizing Cu based electrodes [XX]. In the present study, cells incorporating alloy and ITO electrodes showed same PCE. The higher transmittance of ITO electrode is compensated by the microcavity effect which is strongly prevalent in alloy based device due to its higher reflectance in the visible region (>21 % compared to ~ 9 % for ITO) [XX]. It has been found that the optical resonance effect of the microcavity effect developed between the semi-transparent alloy anode and top Ag cathode leads to enhanced optical electrical field in the ITO free device which ultimately leads to the enhanced optical absorption in the active layer. Beside that lower sheet resistance together with higher work-function of the Ni capped alloy electrodes also plays a crucial role for lowering the injection barrier.

5.6 Conclusions

We have investigated Cu-Ag alloy films as alternative TE structures and demonstrated that they can perform as well as ITO with the advantage of being lower in cost and higher in mechanical flexibility. The optical spectrum of such alloy films follows the average optical behavior of single component Cu and Ag, thus resulting in a much flatter optical response in the visible regime. In addition, the films exhibit very low surface roughness and greater stability in harsh environment when capped with Ni layers. In collaboration with BASF in Germany, we have also shown high efficiency (2.9 %) in OSC devices, equal to those obtained in similar structures using ITO, thus suggesting that indium-free organic devices can be achieved using the proposed alloy based TEs.

Table 5.1 List of electrical characterization data for the OSC devices

	J_{SC}(mA/cm^2)	V_{OC}(mV)	FF(%)	PCE (%)
Alloy	8	770	47	2.9
ITO	7.3	760	51	2.9

References

1. C.G. Granqvist, Sol. Energy Mater. Sol. Cells **91**, 1529 (2007)
2. O. Inganas, Nature Photonics 5, 201, 2011
3. D.S. Ghosh, L. Martinez, S. Giurgola, P. Vergani, V. Pruneri, Opt. Lett. **34**, 325 (2009)
4. K. Sivaramakrishnan, T.L. Alford, Appl. Phys. Lett. **96**, 201109 (2010)
5. J. Lewis, S. Grego, B. Chalamala, E. Vick, D. Temple, Appl. Phys. Lett. **85**, 3450 (2004)
6. M.G. Helander, Z.B. Wang, M.T. Greiner, Z.W. Liu, J. Qiu, Z.H. Lu, Adv. Mater. **22**, 2037 (2010)
7. C.J. Lee, R.B. Pode, J.I. Han, D.G. Moon, Appl. Surf. Sci. **253**, 4249 (2007)
8. S. Giurgola, P. Vergani, V. Pruneri, Nuovo Cimento B **121**, 887–897 (2006)

Chapter 6
Ag/Al:ZnO Nano-Thick Bilayer TEs

6.1 Motivation

The search for an alternate TE other than ITO has led to the development of other TCOs. Al:ZnO (AZO) is another recent choice to be used as one of transparent electrodes (TEs). AZO has advantages like low cost and non-toxicity [1–4]. Another way of getting a more transparent TE is to use the dielectric/metal/dielectric sandwich structure [2, 5–8]. But this kind of triple or more layer structure leads to complexity as one need to deposit the thicker oxide layer in relatively high temperature before and after the metal deposition and also increase the total film-thickness, which is a shortcoming for flexibility.

In this Chapter, we propose a bilayer structure of metal/oxide (Al:ZnO capped Ag nano-thick film) which can overcome the high reflection of metal and retains the good electrical behavior of metal, while keeping the minimum total film-thickness. In this work, AZO and Ag thin films were deposited onto UV grade fused silica substrates and plastic PET substrates by simple DC sputtering at room temperature. Electrical and optical properties of the bilayer structure were theoretically and experimentally studied with variation of Ag thickness. Environmental stability of the proposed bilayer was thoroughly investigated. Figure-of-merit (ϕ_{TE}) was also determined for the bilayer structure and the optimum geometry of the bilayer was applied to the plastic PET substrates, where the mechanical flexibility was investigated and compared with conventional ITO. An additional ultrathin Ni was then added to the bilayer structure to build and OSC which performed similarly to those incorporating ITO.

6.2 Implementation

The bilayer structure of AZO capped Ag nano-thick film (Ag/AZO) was prepared onto 1 mm-thick silica substrates by means of magnetron sputtering (ATC Orion 8 HV). Ag was deposited in pure Ar atmosphere while AZO (aluminum 3 at %

D. S. Ghosh, *Ultrathin Metal Transparent Electrodes for the Optoelectronics Industry*, Springer Theses, DOI: 10.1007/978-3-319-00348-1_6,

doped) was deposited in a mixture of Ar/O_2 (flux ratio 9:1) at room temperature. Deposition rates were determined from the MCM thickness monitor and cross-sectional SEM observation. For a brevity, we denote the different samples with different layer thickness of Ag and AZO as: Ag($\delta1$)/AZO($\delta2$), in which $\delta1$ and $\delta2$ represent the respective thickness of Ag and AZO. The deposition of Ag and AZO was done in one single run without breaking the vacuum. A Perkin Elmer lambda 950 spectrophotometer was used for optical transmission measurements while Cascade Microtech 44/7 S 2749 four-point probe system with a Keithley 2001 multimeter for electrical sheet resistance (R_S) measurements. The morphology of the fabricated films was characterized with the digital instrument D3100-AFM and FEI-SEM. A humidity chamber Vötsch VCL 7003 was used for the damp heat (DH) test in the standard conditions with 85 % relative humidity and at 85 °C. After adding by sputtering an ultrathin Ni layer, the developed TEs were used as anode electrodes in OSCs with the following structure: Anode/MoO_3(5 nm)/ID345:C60(1:1,35 nm)/C60(25 nm)/Bphen(5 nm)/Ag(100 nm).

6.3 Experimental Results and Analysis

Sahu et al. [2] reported nanoscaled Ag on the top of the AZO layer with an attempt to increase the conductivity, in which the AZO thickness was fixed at 25 nm while Ag thickness varied from 4 to 13 nm. It was found that the nanoscaled capping Ag could apparently reduce the optical transparency, with a maximum transparency of 66 % recorded. It is worth noting that the nanoscaled Ag ultrathin films are very unstable and begin to degrade rapidly in the air. In an effort, to stabilize the high conductive Ag layers and improve their optical properties, we covered the Ag layer with an AZO layer. In order to obtain the optimal thickness for the AZO capping layer in the bilayer structure, we carried out calculations based on multiple reflection and refraction as discussed in Chap. 4, Sect. 4.3. Figure 6.1 shows the calculated optical transparency of the Ag/AZO bilayer structure, in which both the thickness of Ag and AZO layer were varied to get the optimum structure. It is seen that the optical transparency of the Ag layer can be greatly improved by a thin AZO capping layer. The maximum transparency peak shows an apparent red-shift with the thickness variation in the capping AZO layers. It is observed that the optimal thickness for the AZO capping layer is in the range of 30–40 nm, in which the average visible optical transparency exhibits the maximum value.

Figure 6.2 compares the performance of Ag/AZO bilayer structure alongside ITO (100 nm), Ag nanowires, SWCNT and graphene film, all of them considered alternative TCs of TCOs. It is observed that the conductivity behavior of the Ag/AZO bilayer structure is mainly dominated by the underlying ultrathin Ag layer when it becomes continuous (thickness >7 nm), for example, if we take a look at the sample properties of Ag(11) and Ag(11)/AZO(30), the R_S for the two samples shows very close values, but for the latter sample, the average visible-optical-transparency is noticeably enhanced from previously 60.5–86.83 %. The transparency spectra for

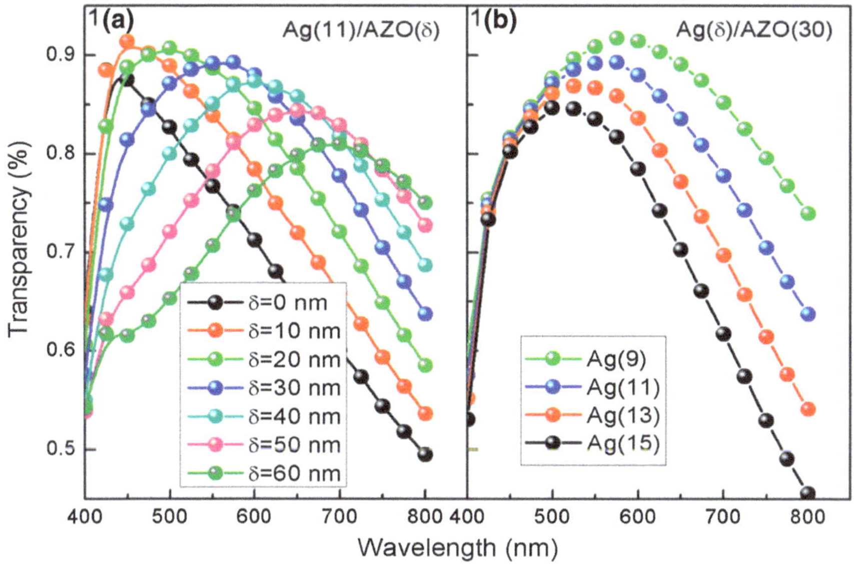

Fig. 6.1 Transparency spectra of **a** Ag(11)/AZO(δ) and **b** Ag(δ)/AZO(30) bilayer structure with different thicknesses (δ) of AZO capping layer and Ag layer respectively, which indicates that the optimal thickness for AZO is in the range of 30–40 nm

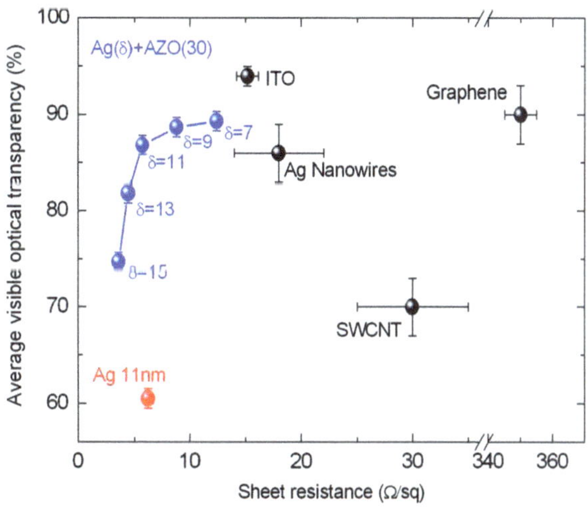

Fig. 6.2 Visible optical transparency, averaged over the 375–700 nm range, as a function of R_S for Ag(δ)/AZO(30) and Ag(11). The value of Ag thickness (δ) identifies each point in the curves. The data for ITO, Ag nanowires, SWCNT and graphene are also shown for a comparison

ITO, Ag (11)/AZO(30) and Ag(11) in the visible light region is shown as the inset in Fig. 6.3. It is seen that the spectrum of Ag (11)/AZO(30) shows a broaden bandwidth and an apparent red-shift with respect to that of Ag(11), which exhibits reasonable consistent with the calculation.

As mentioned earlier (Chap. 4), The Φ_{TE} is conventionally used to evaluate the performance of transparent conductors, where the R_S and optical transparency are

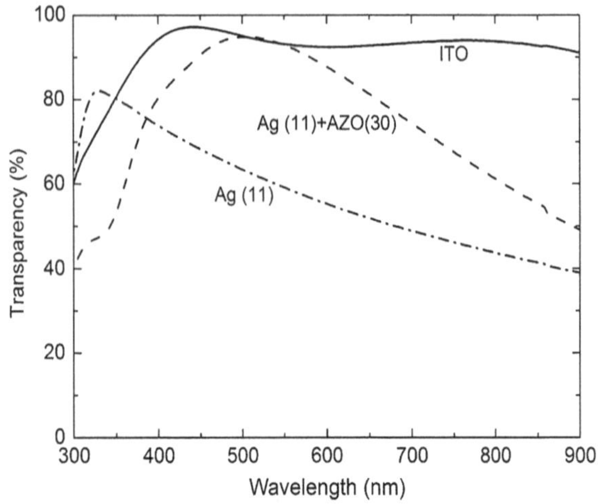

Fig. 6.3 Transparency spectra for ITO, Ag (11)/AZO(30) and Ag(11) in the visible light region

both considered and evaluated. More recently, researchers have used the ratio of DC conductivity and optical conductivity (σ_{DC}/σ_{OP}) as an effective way to compare the various TEs in terms of their performance [9]. The ratio is related to transparency and R_S as $T = (1 + (Z_0/2R_S)(\sigma_{DC}/\sigma_{OP}))^{-2}$, where Z_0 is the impedance of the free space and has the value of 376.8 Ω. High values of σ_{DC}/σ_{OP} implies high transparency and low R_S. We took the average transparency in visible range (375–700 nm) rather than taking it for a particular wavelength for all samples for better evaluation of ϕ_{TE}. For a TE having an average transparency >90 % and $R_S < 10$ Ω/sq, the value of ϕ_{TE} is ≈350, which can be viewed as the benchmark for TE applications. Figure 6.4 shows the calculated ϕ_{TE} values for Ag/AZO bilayer structure alongside ITO (100 nm), Ag nanowires, SWCNT and graphene film. It is seen that the Ag(11)/AZO(30) exhibits the maximum ϕ_{TE} value of about 450, which is much better than ITO and Ag nanowires. Due to the relatively high R_S, graphene and CNT both show very low ϕ_{TE} values. Guillen et. al. [10] achieved the highest ϕ_{TE} in a sandwich ITO/Ag/ITO structure by using Ag film thickness of 10 nm, which gives a good agreement with our experimental result.

The environmental stability of the prepared TEs based on Ag(δ)/AZO(30) was also evaluated using both furnace annealing at an elevated temperature of 500 °C and 7-day DH test with the standard humidity conditions of 85 °C and 85 % relative humidity. In Fig. 6.5, it is clearly observed that annealing process can lead to slight improvement in both electrical and optical properties (dashed line), which can be explained in terms of crystallinity and microstructure enhancement with raised temperature [11]. It is found that the R_S improvement shows saturation behavior after about 300 °C, as can be typically represented by the sample of Ag(11)/AZO(30) in Fig. 6.5 inset. The standard DH test also shows the relatively high environmental stability of our TEs. The samples were kept at 85 % relative humidity and at 85 °C and their electrical and optical properties were measured every 24 h. It is found that the Ag(11)/AZO(30) sample can maintain their opto-electrical properties after the DH test for 1 week. The stability can be evaluated by

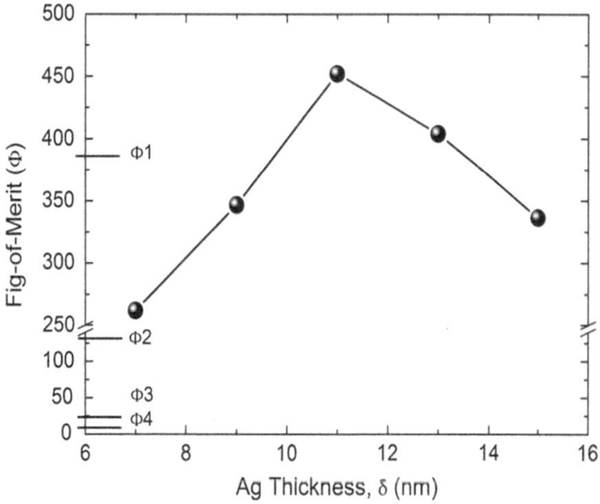

Fig. 6.4 Figure-of-merit for Ag(δ)/AZO(30). Inset is $R_s\delta^3$ versus δ for determination of percolation threshold. Φ_1 to Φ_4 identify the figure-of-merit values for ITO, Ag nanowires, SWCNT, and graphene, respectively

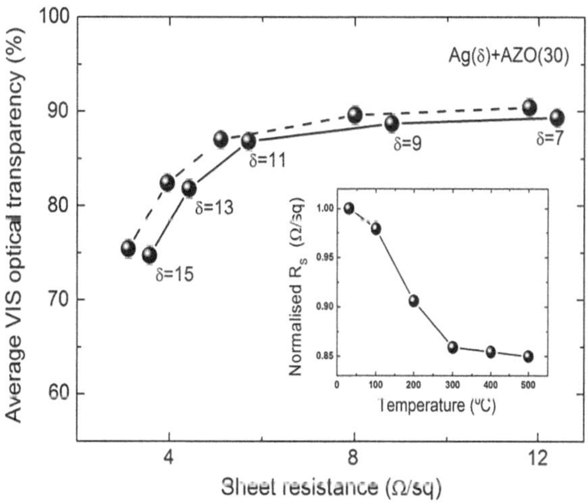

Fig. 6.5 Opto-electrical improvement of Ag(δ)/AZO(30) samples after furnace annealing at an elevated temperature of 500 °C for a dwelling time of 60 min. Inset figure shows the R_S evolution with furnace annealing for Ag(11)/AZO(30)

the relative slope variation in R_S as already mentioned in Chap. 4, Sect. 4.3. One can readily assess a high stability of the Ag(11)/AZO(30) sample (Fig. 6.6). Afterwards, its surface morphology deteriorates with emerging of some droplet-like spots with random distribution on the sample circumference originating from the edges. This is due to the penetration of the water vapor from the unprotected edges of the sample at elevated temperature, since it is well known that the pure Ag ultrathin films are very sensitive to atmosphere, and even degrade rapidly when just exposed to ambient atmosphere.

As the electrode in device application, the transparent conducting layer should have a flat and uniform surface morphology. We characterized the samples prepared on glass and PET with the help of AFM. Ag(11)/AZO(30) samples on glass and PET shows RMS roughness of 2.47 and 2.68 nm respectively over an area of 1 μm square (Fig. 6.7), which evidences the flatness and uniformity of the surface.

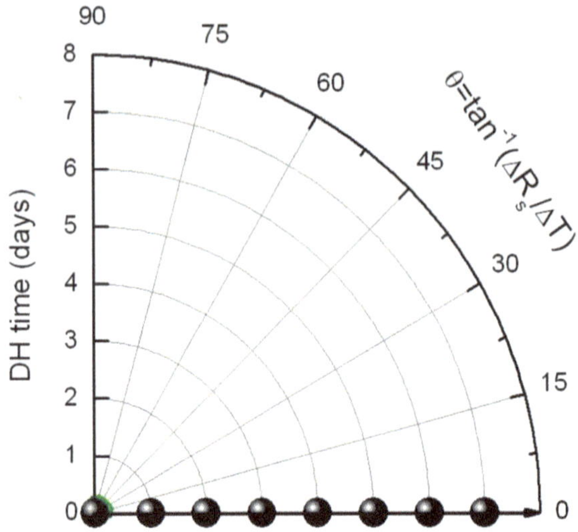

Fig. 6.6 R_S evolution with the standard DH test for Ag(11)/AZO(30)

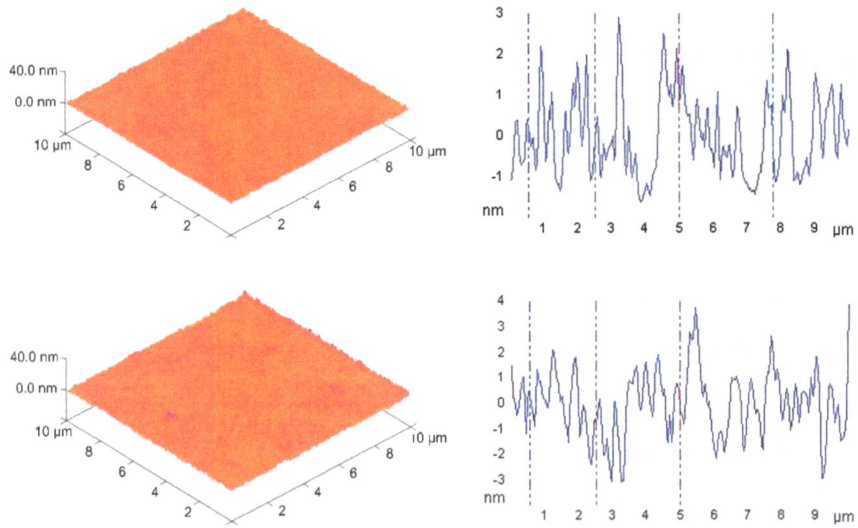

Fig. 6.7 Three dimensional AFM and thickness profile images of Ag(11)/AZO(30) samples on glass (*top*) and PET (*bottom*)

Flexible electronics, such as thin film transistors, flexible solar cells, flexible OLEDs, printable electronics etc., has attracted intense investigation due to the merits like low cost, light weight and superior flexibility [12, 13]. As an indispensible component in flexible electronics, it has become extremely important for TE to retain their electrical and optical properties after undergoing multiple

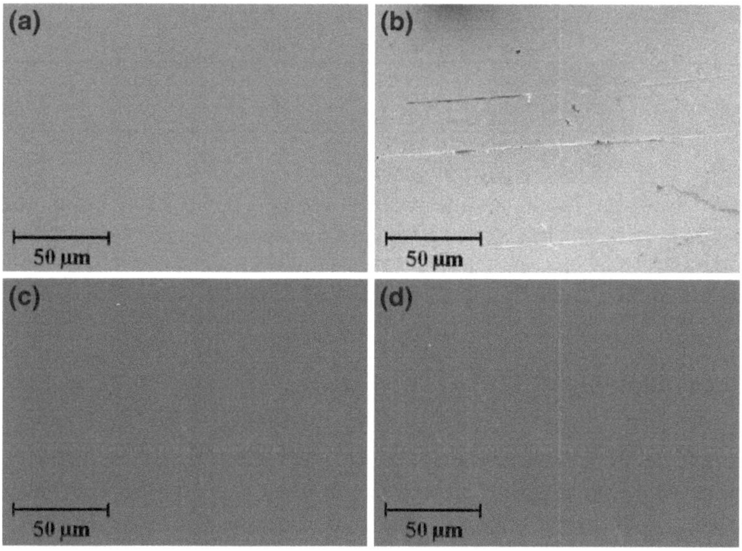

Fig. 6.8 SEM images of the ITO and Ag(11)/AZO(30) films before and after bending. **a** ITO, before bending; **b** ITO, after 20 bending cycles; **c** Ag(11)/AZO(30), before bending; and **d** Ag(11)/AZO(30), after 10^4 bending cycles

bending. To evaluate the flexibility of the proposed Ag/AZO bilayer structure we fabricated several samples on PET substrates with the best film geometry of Ag(11)/AZO(30) and put them to undergo tensile bending from a radius of 12 mm to 4 mm as described in Chap. 5, Sect. 5.3. The results were then compared to state-of-art 100 nm-thick ITO which was commercially bought. It has been observed that the electrical properties of Ag(11)/AZO(30) are very stable even after 10^4 bending cycles while the R_S of ITO increases to 1580 % after 20 such cycles as mentioned in Chap. 4. In the case of Ag(11)/AZO(30), ductile Ag conductive layer beneath the AZO in our proposed structure helps in retaining the electrical properties after bending [14]. Figure 6.8 shows the SEM images of the ITO and Ag(11)/AZO(30) films before and after bending test. Cracks are formed in the case of ITO while no observable degradation in surface morphology is found in the case of Ag(11)/AZO(30).

6.4 Performance in an OPV Device

The best structure Ag(11)/AZO(30) were then used to fabricate OSCs. 1 nm-thick nickel is added to enhance the work function and stability of the Ag/AZO anode [15]. The electrical characterization data comparing the performance of devices incorporating Ag/AZO/Ni and ITO anodes are reported in Table 6.1. The values for open circuit voltage (V_{OC}) and fill factor (FF) for Ag/AZO/Ni and ITO devices are

Table 6.1 List of electrical characterization data for the OPV device

	Voc (mV)	Jsc (mA/cm^2)	FF (%)	PCE (%)
ITO	760	7.3	53	2.9
Ag(11)/AZO(30)	760	6.6	53	2.6

practically the same. The main difference between the devices is in the short-circuit current density (J_{SC}). However, the device incorporating an Ag/AZO/Ni anode shows about 90 % of the efficiency of that incorporating ITO.

6.5 Conclusions

Transparent electrode structures made of Ag/AZO films on glass or PET substrates were prepared by DC magnetron sputtering at room temperature. The thickness of the Ag and AZO layer is optimized to maximize the antireflection effect. It is found that 30 nm thick AZO on the top of Ag can greatly enhance the optical transparency. The dependence of electrical and optical properties of the bilayer Ag/AZO structure as a function of Ag thickness is studied. It is found that Ag(11)/AZO(30) has a figure-of-merit larger than other TEs, such as ITO, single wall carbon nanotubes, graphene and Ag nanowires. The proposed electrodes are stable against elevated temperature (500 °C) and damp heat conditions (85 °C and 85 % relative humidity). After adding Ni to increase the work function, the Ag/AZO/Ni electrode on PET was used as an anode in an organic solar cell achieving more than 90 % of the efficiency of a similar device incorporating ITO. The R_S of the proposed electrode remains unchanged after 10^4 bending cycles while its surface does not show any sign of cracking after 10^4 bending cycles. On the contrary ITO on PET shows an exponential increase in R_S and surface cracking only after 20 cycles. Considering also its low cost and the room temperature processing, the developed TEs have great potential for the flexible optoelectronics industry.

References

1. W. Tang, D.C. Cameron, Aluminum doped zinc oxide transparent conductors deposited by the sol-gel process. Thin Solid Films **238**, 83 (1994)
2. D.R. Sahu, J.Y. Huang, Development of ZnO based transparent conducting coatings. Sol. Energy Mater. Sol. Cells **93**, 1923 (2009)
3. S. Fernandez, F.B. Naranjo, Optimization of aluminum doped zinc oxide films deposited at low temperature by radio frequency sputtering on flexible substrates for solar cell applications. Thin Solid Films **94**, 157 (2010)
4. P.C. Lansaker, J. Backholm, G.A. Niklasson, C.G. Granqvist, TiO2/Au/TiO2 multilayer thin films: novel metal based transparent conductors for electrochromic devices. Thin Solid Films **518**, 1225 (2009)

5. L. Cattin, M. Morsli, F. Dahou, S.Y. Abe, A. Khelil, J.C. Bernede, Investigation of low resistance transparent MoO3/Ag/MoO3 multilayer and application as anode in organic solar cells. Thin Solid Films **518**, 4560 (2010)
6. X. Lui, X. Cai, J. Qiao, J. Mao, N. Jiang, The design of ZnS/Ag/ZnS transparent conductive multilayer films. Thin Solid Films. **441**, 200 (2003)
7. D.R. Sahu, S.Y. Lin, J.Y. Huang, Deposition of Ag-based Al-doped ZnO multilayer coatings for the transparent conductive eletrodes by electron beam evaporation. Sol. Energy Mater. Sol. Cells **91**, 851 (2007)
8. G.B. Murdoch, S. Hinds, E.H. Sargent, S.W. Tsang, L. Mordoukhovski, Z.H. Lu, Aluminum doped zinc oxide for organic photovoltaics. Appl. Phys. Lett. **94**, 213301 (2009)
9. A. Ojala, H. Bürckstümmer, M. Stolte, R. Sens, H. Reichelt, P. Erk, J. Hwang, D. Hertel, K. Meerholz, F. Würthner, Parallel bulk-heterojunction solar cell by electrostatically driven phase separation. Adv. Mater. **23**, 5398 (2011)
10. C. Guillen, J. Herrero, ITO/Metal/ITO multilayer structures based on Ag and Cu metal films for high-performance transparent electrodes. Sol. Energy Mater. Sol. Cells **92**, 938 (2008)
11. M. Neghabi, A. Behjat, S.M.B. Ghorashi, S.M.A. Salehi, The effect of annealing on structural, electrical and optical properties of nanostructured ZnS/Ag/ZnS films (2011). doi: 10.1016/j.tsf.2011.03.023
12. K. Nomura, H. Ohta, A. Takagi, T. Kamiya, M. Hirano, H. Hosono, Room temperature fabrication of transparent flexible thin-film transistors using amorphous oxide semiconductors. Nature **432**, 488 (2004)
13. P. Heremans, Organic crystals at large. Nature **444**, 828 (2006)
14. B. Sim, E.H. Kim, J. Park, M. Lee, Highly enhanced mechanical stability of indium tin oxide film with thin Al buffer layer deposited on plastic substrate. Surf. Coat. Technol. **204**, 309 (2009)
15. J.C. Bernède, Y. Berredjem, L. Cattin, M. Morsli, Improvement of organic solar cell performances using a zinc oxide anode coated by an ultrathin metallic layer. Appl. Phys. Lett. **92**, 83304 (2008)

Chapter 7
Ultrathin Metal Capped Al-Doped ZnO as Transparent Electrode

7.1 Motivation

The rising cost and problems associated with ITO has led to a search for alternate transparent electrode (TE) material. Other TCO materials are being widely developed and in particular Al-doped ZnO (AZO) has received great attention in the recent years and has been considered as a promising alternative to ITO [1–4]. This material is particularly interesting because of its low cost and wide availability of its constituent raw materials. However, the environmental stability of AZO films is still an unsolved problem, leading to significantly increased electrical resistivity and surface degradation when the AZO is exposed to harsh environment (e.g., annealing in air or humidity damping). This drawback prevents AZO films to be deployed in real applications that require long-term reliability of the devices. Therefore, great effort has been dedicated to improve the environmental stability of AZO thin films [3–6]. Miyata et al. [3] pointed out that oxygen might be responsible for the degradation of electrical properties during thermal annealing in air. Greiner et al. [7] showed that the surface morphology plays an important role in AZO environmental stability, while Lin et al. [4] observed a degradation in electrical conductivity, but negligible change in optical properties. Up to now, the majority of the research works have targeted composition modification as well as crystalline quality control to stabilize AZO [4–7].

The present work aims to provide the AZO electrodes with more transparency, stability, protection and compatibility with the environment. In this chapter, we investigate the environmental stability of AZO thin films and report an effective approach to enhance it. By adding an ultrathin Ni capping layer with a thickness in proximity of its percolation threshold (2.5 nm thickness), the environmental stability of AZO thin film is greatly enhanced, as it is shown via damp heat (DH) tests at 95 °C and 95 % relative humidity. The developed Ni capped AZO TE is then used as an anode in OLEDs and OSCs, showing performance similar to those of ITO-based devices.

D. S. Ghosh, *Ultrathin Metal Transparent Electrodes for the Optoelectronics Industry*, Springer Theses, DOI: 10.1007/978-3-319-00348-1_7,

7.2 Implementation

AZO thin films with thicknesses of 30 and 250 nm, named as AZO_S and AZO_L, respectively, and Ni layer with different thicknesses were deposited by sputtering on optically polished UV fused silica substrates. Prior to the deposition, the substrates were cleaned in standard way as mentioned in Chap. 2, Sect. 2.2. The sputtering target used was Zinc Oxide target 3 at % doped with Al. AZO films were prepared with a substrate temperature of 200° C, RF power 150 W, and a pure Ar working pressure of 1–1.5 mTorr. Nickel was deposited at 100 W DC power in pure Ar atmosphere of 1.5 mTorr. The electrical sheet resistance (R_S) was measured by Four point Probe (Cascade Microtech 44/7S) while for the transmission spectra, Perkin Elmer Lambda 950 spectrophotometer was used. A humidity chamber Votsch VCL 7003 was used for the DH test in harsh conditions of 95° C and 95 % relative humidity. The surface morphology was characterized by atomic-force-microscope (AFM) and optical interferometry (Veeco Wyko 9800NT). Work function of the films was evaluated by ambient scanning Kelvin probe microscopy with gold as a reference electrode. Conventional X-ray diffraction (XRD) and X-ray photoelectron spectroscopy (XPS) were used for identifying crystal structure and elemental composition, respectively.

7.3 Experimental Results and Stability Analysis

There are already some reports on the combination of metal and TCOs as TEs. Recently, Bernede et al. deposited a very thin metal layer (0.5–1.5 nm), preferably 0.5 nm, on the top of TCO to improve their functionality [8]. It is found that such an UTMF improves the device performance due to the better matching of energy levels between the transparent electrode and organic layer which in turn implies lower injection barrier. However, in the case of capping the TCO with a metal layer, too thick metal capping layer would lead to large light absorption while too thin metal layer would be prone to formation of discrete island-like morphology, thus resulting in an incomplete protection of the underlying AZO against the environment or detrimental interaction with the other layers forming the device. Therefore, it is intuitive that the optimum thickness of the metal capping layer should be around its percolation threshold, i.e., the thickness corresponding to which the layer morphology changes from an island distribution to a continuous layer. The percolation threshold can be estimated by plotting the $R_S t^3$ versus t as also described in Chap. 4, Sect. 4.3. Figure 7.1a shows this plot for Ni layer, indicating a percolation threshold between 2 and 3 nm. Another way of inferring the percolation threshold is to use the F–S model (described in Chap. 2, Sect. 2.5.1), which for a continuous and uniform thin film defines the relationship between thickness (t) and resistivity (ρ_f). Figure 2.7 in Chap. 2 shows the experimental data for Ni films fitted with F–S model. From the fitting curve, it is observed that the F–S model can predict the

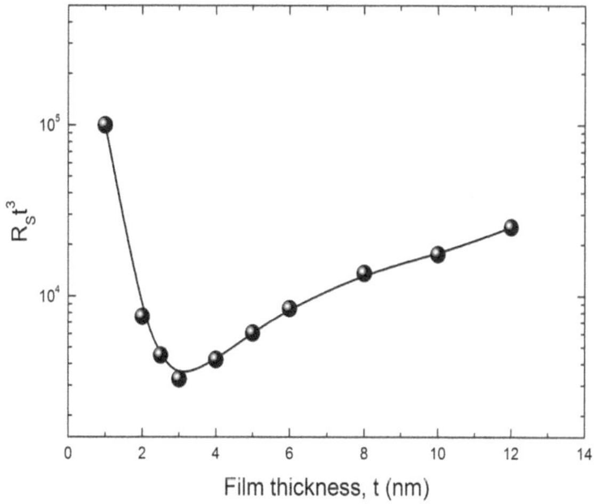

Fig. 7.1 $R_S t^3$ versus *t* for the determination of percolation threshold

resistivity of Ni layer very well up to approximately 2.5 nm. The experimental resistivity value of 2 nm-thick Ni shows a clear deviation from the corresponding fitting data, indicating that a Ni layer with a thickness less than 2.5 nm is no longer continuous and uniform. It is interesting to note that the resistivity evolution of Ni layers with varied thickness on insulating ZnO/glass shows the same trend, indicating that the percolation threshold for the room temperature deposited Ni layers on glass can be applicable to AZO/glass. The aforesaid analysis indicates that the percolation threshold for Ni layer (t_P) is about 2.5 nm. For a thickness less than t_P island-like structures will form. We thus name Ni_P and Ni_I for the Ni layers with thicknesses t_P and less than t_P, respectively.

It is worth noticing that the optical transmittance of Ni layers depends strongly on their thickness, as shown in the Fig. 7.2. In order to recover the transparency,

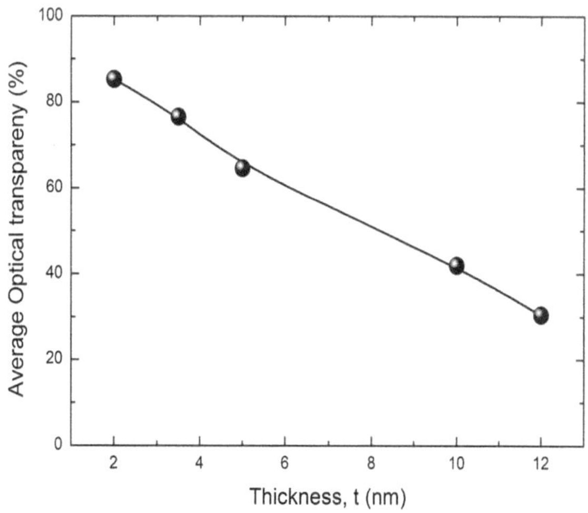

Fig. 7.2 Average optical transparency for Ni films with varied thickness

all the AZO/Ni samples were then in situ treated in O_2 plasma. Apart from the transparency recovery, the O_2 plasma also has the role of enhancing the surface work function through the Ni oxide formation [9]. The resulting work function values were measured 4.75 and 5.34 eV for AZO and AZO/Ni_TR (where TR denotes treatment with O_2 plasma), respectively, thus confirming a significant enhancement.

In order to investigate the transparency of AZO/Ni samples, we calculated the transmission of our films based on a multilayer stacked structure by solving the Fresnel equations, the matrix formalism described in Chap. 4, Sect. 4.3. Figure 7.3: (a) Theoretical calculation of the three layer AZO/Ni/NiOstacks on fused silica substrates alongside with the experimental data of pure AZO, AZO/Ni(2.5), and AZO/Ni(2.5) _TR, (b) XPS spectrum of the structure AZO/Ni(2.5)_TR showing the presence of metallic Ni. Figure 7.3 (a) shows the results of the theoretical calculation for the three layer AZO/Ni/NiO structure on fused silica glass substrate, alongside with the experimental data of AZO, AZO/Ni(2.5)_TR, and AZO/Ni(2.5). There is a good agreement between the calculated and experimental data for the structure. As it is predicted by the calculation, the formation of Ni oxides can effectively raise the transparency of AZO/Ni(2.5) samples. It is pointed out that the presence of the remaining Ni, even though in minimum quantity is essential for ensuring that the oxidation treatment will not influence the interface between AZO and Ni. The O_2 plasma treatment was thus optimized so that the AZO/Ni(2.5)_TR samples can maintain the interface integrity while exhibiting high transparency. From Fig. 7.3a, it is seen that the calculated curve of AZO/Ni(0.5)/NiO_X(2) is close to the experimental data of AZO/Ni(2.5)_TR, thus indicating that the remaining Ni thickness (0.5 nm) is close to that of the unit-cell dimension (about 0.35 nm). The X-ray photoelectron spectra (XPS) analysis of our samples have also shown the presence of metallic Ni (852.9 eV) on the surface of AZO/Ni(2.5)_TR, after removal of the top layer of oxidized Ni by Ar^+ sputtering (Fig. 7.3b).

For the experiments, we fabricated and measured the following samples: AZO_L (250 nm), AZO_L/Ni_P (250 nm/2.5 nm-oxidized). Figure 7.4 shows the DH test results for AZO_L and AZO_L/Ni_P samples. It is clearly seen that AZO_L/Ni_P sample shows electrical properties more stable than those of AZO_L, with a sheet resistance value still below 24 Ω/sq after 30 days (Fig. 7.4a). Moreover, as it can be seen in the Fig. 7.4b, the average optical transmittance in the 375–700 nm wavelength range for AZO_L/Ni_P remains unchanged—in fact it even increases slightly due to additional oxidation of the remaining Ni while that of AZO_L strongly decreases. The surface morphology of AZO_L and AZO_L/Ni_P samples was also characterized by optical interference microscopy. Before the DH test, a typical image for both samples is featureless (Fig. 7.5a). After 30-day DH, the surface morphology of AZO_L/Ni_P remains almost unchanged while that of AZO_L shows spikes and droplet-like spots distributed in a random manner (Fig. 7.5b). The morphological degradation of AZO_L is likely to be intimately correlated with hydrolysis-induced corrosion process occurred possibly at the grain boundaries. The root-mean square (RMS) roughness values for as-deposited AZO_L and AZO_L/Ni_P are about 3.18 and

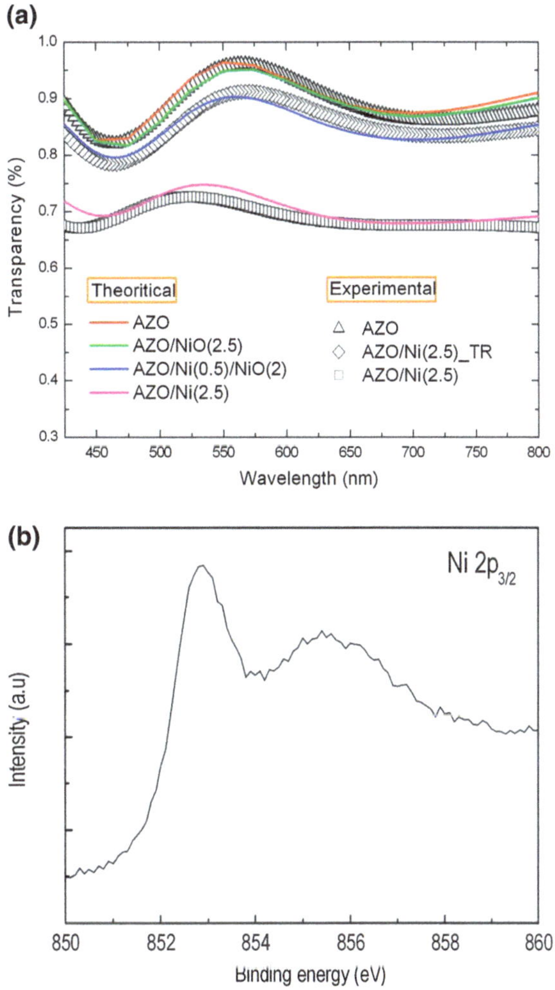

Fig. 7.3 **a** Theoretical calculation of the three layer AZO/Ni/NiO_X stacks on fused silica substrates alongside with the experimental data of pure AZO, AZO/Ni(2.5), and AZO/Ni(2.5) _TR, **b** XPS spectrum of the structure AZO/Ni(2.5)_TR showing the presence of metallic Ni

2.33 nm, respectively. Furthermore, the surface morphology of AZO_L and AZO_L/Ni_P samples after the DH test was also characterized by AFM, as it is shown in Fig. 7.4c and d. The AZO_L sample shows a very rough surface with a RMS roughness value greater than 100 nm while the AZO_L/Ni_P sample exhibits an almost unchanged surface roughness compared to as deposited value. This indicates that deposited Ni capping can provide a dense, uniform, and smoothing protective layer to avoid any surface degradation during DH test. From the above results, one can conclude that oxidized Ni capping layers with appropriate thickness can effectively increase the environmental stability of AZO films and strongly reduce otherwise unavoidable degradation in surface morphology, electrical, and optical properties.

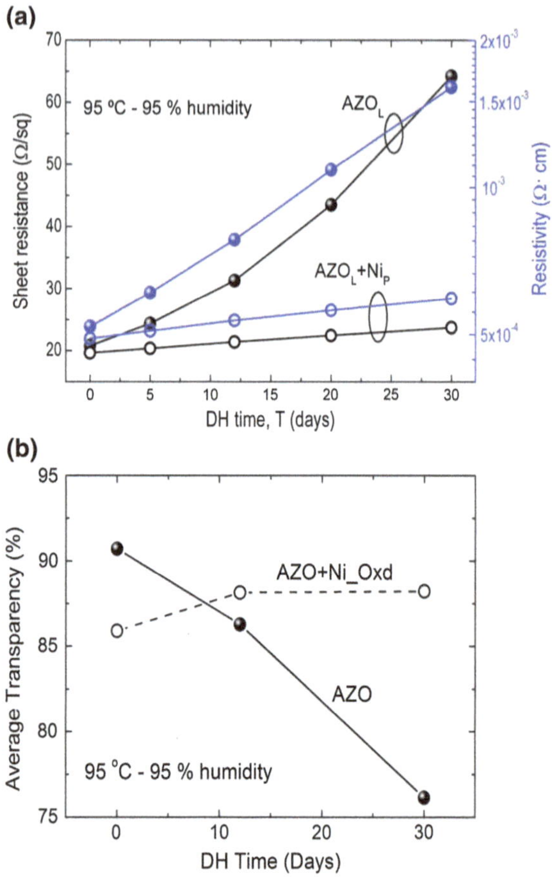

Fig. 7.4 Correlation of properties of AZO_L and AZO_L/Ni_P films with DH test time **a** electrical **b** average optical transmittance

Miyata et al. [3] reported that AZO layers with a thickness below 50 nm were very unstable even when exposed to air at room temperature for a long time. The adsorption of oxygen at grain boundaries was found to be responsible for the degradation. Therefore, to further elucidate the role of Ni_P, environmental stability tests of AZO_S, AZO_S/Ni_I, and AZO_S/Ni_P samples, where the thickness of AZO is approximately 30 nm, Ni_P is 2.5 nm, Ni_I is 1 nm, was investigated under both thermal annealing and DH conditions. The stability after thermal annealing can be evaluated by the relative slope variation in sheet resistance with $\theta = 0$ corresponds to ideal stability. One can readily assess a much higher stability of AZO_S/Ni_P over AZO_S and AZO_S/Ni_I samples (Fig. 7.6b). Ni_I capping layer cannot cover the entire surface and thus AZO_S/Ni_I shows an inferior stability, but due to the presence of the Ni island formation, the stability in this case is still better than AZO_S. The DH test further evidences the best stability for AZO_S/Ni_P sample (Fig. 7.6a).

From Fig. 7.7 one can see that the AZO film grown on glass shows a textured structure with a very strong preferential alignment of AZO(002), in which the full width at half maximum (FWHM) for AZO(002) is only 0.34°. The average grain

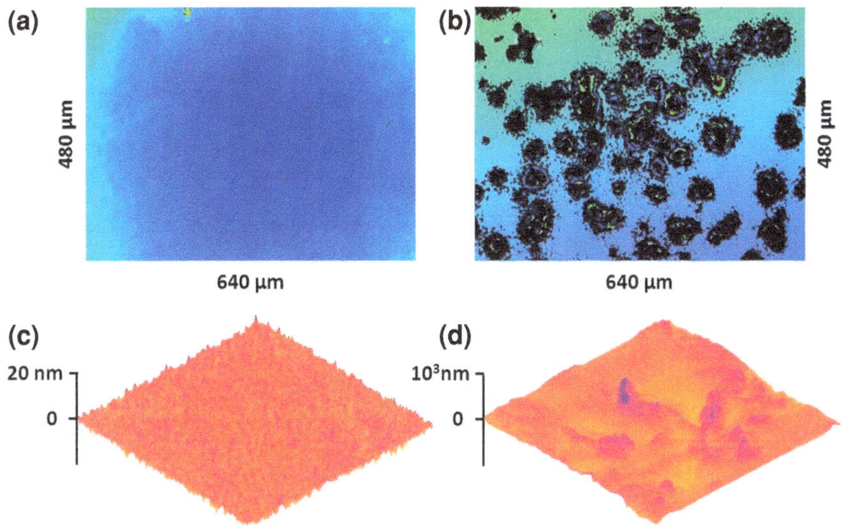

Fig. 7.5 **a** Typical optical micro image for AZO_L and AZO_L/Ni_P samples before DH test and **b** for AZO_L after a 30 days DH test, **c** and **d** are the AFM images (scanning area of $10 \times 10\ \mu m^2$) for AZO_L/Ni_P and AZO_L after the DH test respectively

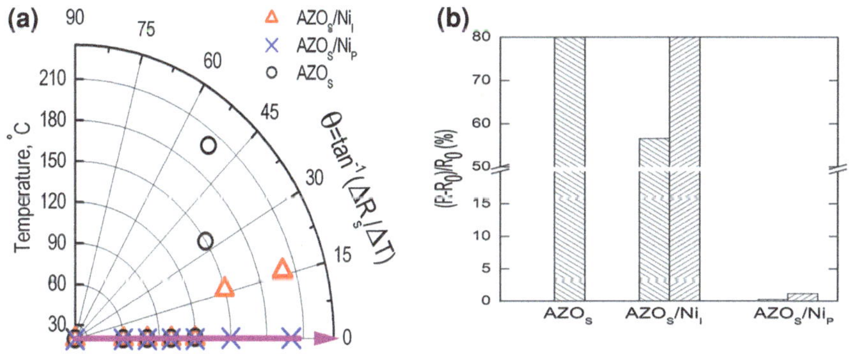

Fig. 7.6 **a** Polar figure for the thermal stability of the samples, the horizontal axis ($\theta = 0$) corresponds to ideal stability and **b** sheet resistance changes with DH test time (*red column* 24 h and *green column* 1 week) for different samples

size is calculated to be around 35 nm using Scherrer formula. The calculated grain size is much higher than reported in literature confirming the better quality of deposited AZO [10]. The degree of out-of-plane alignment is evaluated by a ω scan (rocking curve for the AZO (002) peak), as it is shown in the inset of Fig. 7.7, the FWHM for it being 4.8°. It is believed that the textured crystalline structure of our AZO films plays a decisive role in enhancing their environmental stability.

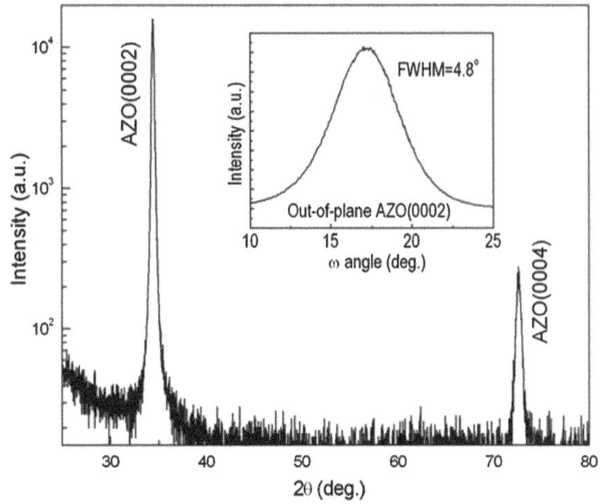

Fig. 7.7 XRD profile for AZO grown on glass showing highly crystalline structure, inset is the scan for the AZO (002) peak

7.4 Performance in OLED and OPV Device

In collaboration with Stephanie Cheylan and Danny Krautz (ICFO), as deposited AZO and AZO_L/Ni_P was also used, instead of ITO, as a transparent hole-injecting electrode for bottom polymer LEDs. Standard architecture of the diodes used for the study was: Anode/PEDOT:PSS/PFO/CaAg. Poly(phenylenevinylene) copolymer (SY), commercially available from Merck, was dissolved in toluene, stirred overnight and filtered before being spin-coated onto the PEDOT-PSS layer for a thickness of around 80 nm. Ca and Ag are sequentially thermally evaporated ($<5 \times 10^{-6}$ Torr) onto the emitting layer using shadow mask (Table 7.1).

For a comparison, commercially available ITO (100 nm and 17–21 Ω/sq) was used to realize the same device architecture. The AZO_L/Ni_P anode based device shows an equivalent performance to the ITO-based one. The enhanced device performance based on AZO_L/Ni_P anode is related to a significant improvement in the injection of charges due to a better work function matching at the anode-organic interface.

In addition, in collaboration with the Organic Nanostructured Photovoltaics group in ICFO, the developed TE was also used to make OSCs. It was recently

Table 7.1 Performance data of the devices with ITO and AZO electrodes. V_{TH} corresponds to the voltage at which 1 cd m^{-2} is obtained and V_{maxLce} corresponds to the voltage at which the maximum luminance is obtained

	Work function (eV)	Current density (A m^{-2})	V_{TH}/VmaxLce (V)	Luminance (cd m^{-2})
AZO_L	4.75	7930	6/23	120
AZO_L/Ni_P	5.3	7930	4.6/11.2	145
ITO	5.0	7930	4.6/10.2	150

Table 7.2 List of electrical characterization data for the OPV device

	J_{SC} (mA/cm^2)	V_{OC} (mV)	FF (%)	PCE (%)
ITO/NiO_X	−10.62	520	59	3.25
AZO_L/NiO_X	−7.50	485	25	0.91
AZO_L/Ni_P	−8.29	472	60	2.36

shown that NiO_X could be used as hole transporting layer (HTL) for OSCs providing better efficiency and stability [11]. We found that the AZO_L/Ni_P can both act as anodic TE and HTL for bulk-heterostructure OSCs at the same time. The fabricated OSC has the following configuration: TE/P3HT:PCBM/$LiCoO_2$/Al, in which the AZO_L provides the dominant electrical conductivity while the top Ni_P works as an excellent work function matching layer, an electron blocking layer, as well as a protective layer for the underlying AZO (Table 7.2).

The OPV cell based on AZO_L/Ni_P reaches 73 % of the efficiency of an ITO based structure which has an additional NiO_X film on top of it as HTL. As mentioned earlier, there is an ultrathin Ni interlayer remaining after the O_2 plasma treatment for AZO_L/Ni_P. In order to investigate the role of the remaining Ni interlayer, we fabricated an OSC with the following architecture: AZO_L/NiO_X/P3HT:PCBM/$LiCoO_2$/Al, in which the NiO_X layer was sputtered from a NiO sintered ceramic target and deposited on AZO_L. The IV curve shows an S-shaped feature which could be due to the energy barrier for carrier extraction inside the interfaces of OSC. From these studies, it is clear that such a barrier is present between AZO and NiO_X, possibly related to the difference in work function (4.75 eV for AZO eV and 5.4 for NiO). It is expected that the final device performance can be further improved by optimization of the device stack

7.5 Conclusions

In summary, an ultrathin Ni capping layer with a thickness at percolation threshold can significantly stabilize an underlying AZO layer in harsh environment. The underlying AZO dominates the electrical conductivity while the top Ni/NiO works as an excellent work function matching layer, an hole transporting layer, as well as a protective layer for the underlying AZO. The proposed capping layer acts as a blocking layer to inhibit the penetration of oxygen and water at the grain boundaries of AZO. The bilayer transparent electrode shows electro-optical performance similar to ITO when used as an anode in OLED devices and has been effectively used as a versatile transparent electrode acting as both anode and HTL layer in an OPV cell.

The present approach can be extended to other TCOs such as FTO, GZO and TTO for their environmental stability and functional improvement. On the other hand, other metal capping layer such as Ti instead of Ni can also be used which may in turn can act as hole blocking layer (HBL) for organic electronic devices.

The scheme can also be applied with the protective layer is coated both below and top of the TCO. This particular arrangement can be very critical for plastic substrates which absorb a substantial amount of moisture and potentially lead to the degradation of conducting layer.

References

1. T. Minami, Present status of transparent conducting oxide thin-film development for Indium-Tin-Oxide (ITO) substitutes. Thin Solid Films **516**, 5882 (2008)
2. G. Murdoch, S. Hinds, E. Sargent, S. Tsang, L. Mordoukhovski, Z. Lu, Aluminum doped zinc oxide for organic photovoltaics. Appl. Phys. Lett. **94**, 213301 (2009)
3. T. Miyata, Y. Ohtani, T. Kuboi, T. Minami, Stability of nano-thick transparent conducting oxide films for use in a moist environment. Thin Solid Films **516**, 1354 (2008)
4. W. Lin, R. Ma, J. Xue, B. Kang, RF magnetron sputtered ZnO:Al thin films on glass substrates: A study of damp heat stability on their optical and electrical properties. Sol. Energy Mater. Sol. Cells **91**, 1902 (2007)
5. F. Ruske, M. Roczen, K. Lee, M. Wimmer, S. Gall, J. Hupkes, D. Hrunski, B. Rech, Improved electrical transport in Al-doped zinc oxide by thermal treatment. J. Appl. Phys. **107**, 13708 (2010)
6. F.J. Pern, S.H. Glick, X. Li, C. DeHart, T. Gennett, M. Contreas, T. Gessert, Stability of TCO window layers for thin-film CIGS solar cells upon damp heat exposures: part III. Proc. SPIE **7412**, 74120K (2009)
7. D. Greiner, N. Papathanasiou, A. Pflug, F. Ruske, R. Klenk, Influence of damp heat on the optical and electrical properties of Al-doped zinc oxide. Thin Solid Films **517**, 2291 (2009)
8. J.C. Bernede, Y. Berredjem, L. Cattin, M. Morsli, Improvement of organic solar cell performances using a zinc oxide anode coated by an ultrathin metallic layer. Appl. Phys. Lett. **92**, 83304 (2008)
9. M. Chan, T.Y. Hsu, F.C. Hong, Enhanced hole injections in organic light-emitting devices by depositing nickel oxide on indium tin oxide anode. Appl. Phys. Lett. **81**, 1899 (2002)
10. J.H. Park, J.M. Shin, S.Y. Cha, J.W. Park, S.Y. Jeong, H.K. Pak, C.R. Cho, Deposition temperature effects on AZO thin films prepared by RF magnetron sputtering and their physical properties. J. Korean Phys. Soc. **49**, S584 (2006)
11. R. Betancur, M. Maymo, X. Elias, L.T. Vuong, J. Martorell, Sputtered NiO as electron blocking layer in P3HT:PCBM solar cells fabricated in ambient air. Sol. Energy Mater. Sol. Cells **95**, 1228 (2011)

Chapter 8
Summary and Future Outlook

Inexpensive, low temperature processable and flexible TEs are essential for the functionality of energy efficient optoelectronic devices, such as OPV cells, OLEDs, and displays. The main limitations of the currently used TEs, including ITO, are related to their high cost, low compatibility with active organic materials and poor mechanically flexibility. This thesis has investigated the properties and applicability of UTMF based TEs, showing that they are an effective and viable alternative to ITO for the optoelectronics industry.

The main results include:

- The investigation of the optical response from the UV to the mid-infrared region of single component Ni and Cr based UTMFs, showing that the optical transmission of the deposited films is comparable to that of ITO in the visible and near-infrared range (0.4–2.5 μm), while it can be significantly higher in the ultraviolet (175–400 nm) and mid-infrared (2.5–25 μm). This feature makes them ideal TE candidates for devices like UV photodiodes, UV LEDs, solar cells for space satellites, and IR pyroelectric detectors.
- An innovative technique to make the UTMF based TEs environmentally stable. The oxidation mechanism, which is a major concern for metal films, can be in fact exploited to achieve stable films. With an appropriate temperature cycle in the presence of oxygen, a protective oxide layer on top of the metallic film is produced, which prevents further oxidation of the underneath metal.
- An innovative TE structure, incorporating an *ad hoc* conductive grid, which can significantly reduce the sheet resistance of UTMF based transparent electrodes, leaving practically unchanged their transparency. By employing this design, we were able to reduce the sheet resistance of a continuous 2 nm Ni film (>950 Ω/sq) to 6.5 Ω/sq while the transparency was maintained at values exceeding 75 %. The design allows going beyond the current trade-off between optical transparency and electrical conductivity of continuous metal based TEs.

D. S. Ghosh, *Ultrathin Metal Transparent Electrodes for the Optoelectronics Industry*, Springer Theses, DOI: 10.1007/978-3-319-00348-1_8,

- The combination of different UTMFs into the same TE structures, for example, ultrathin Cu films and Cu–Ag alloys capped by a Ti or Ni layer. These TEs show excellent stability against temperature and oxidation and present also the appropriate work function. The performance of Cu–Ni and Ag–Cu alloy-Ni based TEs has been investigated in devices like OLEDs and OSCs, achieving conversion efficiencies around 3 %, comparable to those of devices using ITO.
- The bilayer doped metal oxide/metal (AZO capped Ag nano-thick film) which can overcome the high reflection of metals and retain their good electrical behavior, while still keeping the minimum total film-thickness. Electrical and optical properties as well as the environmental stability of the doped metal oxide/metal bilayer structure were theoretically and experimentally studied. In an attempt to make a hybrid TE, UTMF capping layer were also successfully utilized for stability improvement of AZO. It was found that an ultrathin oxidized Ni capping layer with a thickness at percolation threshold greatly enhances the stability of AZO layer in harsh environment without affecting the electro-optical properties. We have also verified that the combined structure act both as anode and hole transporting layer, the latter property generally achieved by depositing another layer during device fabrication.

Although this thesis has contributed to the foundation of a novel class of transparent electrode materials and showed their effective incorporation into OLEDs and OSCs, there still remain significant developments before UTMFs have a widespread use in optoelectronic applications.

An area of high interest is widening the applications beyond OLEDs and OSCs, including flat-panel rigid displays for TV screens, tactile displays for smart-phone and tablets, and sensors. For this purpose, the performance obtained on glass rigid substrates should also be achieved on flexible plastic sheets, such as PET and PEN, this being crucial for a low cost roll-to-roll production. For example the UTMF combined with the grid concept offers great margin of improvements together with mechanical flexibility by choosing new material combinations.

A disadvantage of using polymeric substrates is that they absorb a substantial amount of moisture which eventually penetrates through it and potentially leads to the degradation of UTMFs. This is a difficult problem to be solved and substantial efforts are needed to develop effective barrier layers, which can enhance the stability as well as the adhesion of the UTMFs without compromising their flexibility.

Besides the aforementioned UTMF with grid structure especially suitable for flexible transparent electrodes, the developed UTMF technology can be combined with other ITO alternatives, including CNT, graphene or metallic nanowires, to achieve improved electro-optical and stability performance. From a material perspective the substrate will also play a significant role in the future. Nano-structuring of the surface could lead to additional properties for the UTMF transparent electrodes deposited on top of it, including hydrophobicity and self-cleaning capabilities.

Curriculum Vitae

DHRITI SUNDAR GHOSH
ICFO-The Institute of Photonic Sciences
Optoelectronics Group—Nanophotonics
Mediterranean Technology Park
Av. Carl Friedrich Gauss, 3
08860 Castelldefels (Barcelona), Spain
E-mail: dhritisghosh@gmail.com / dhriti.ghosh@icfo.es;

Present Designation

Postdoctoral fellow at ICFO-The Institute of Photonic Sciences (www.icfo.eu), Castelldefels, Barcelona, Spain.

PhD Thesis Work

Thesis Advisor: Prof. Valerio Pruneri (**ICFO**)
Thesis Title: Ultrathin metal transparent electrodes for the optoelectronics industry

The objective of my PhD work was to develop single/multiple component ultrathin ($\leq$10 nm) metal film (UTMFs) based Transparent Electrodes (TEs) which can replace expensive ITO, and to integrate the technology into the industrial flow. I studied morphological, electrical, optical, structural and stability properties of UTMFs of different metals and their combinations for potential

D. S. Ghosh, *Ultrathin Metal Transparent Electrodes for the Optoelectronics Industry*, Springer Theses, DOI: 10.1007/978-3-319-00348-1,

applications in OLEDs, photovoltaic cells (organic and inorganic both), displays, electrochromic and modulation devices.

My work also explored the possibility of combining UTMFs with indium-free TCOs, Carbon Nanotubes, Graphene and 1-D nanostructures for hybrid TEs.

In addition, my work has led to the collaboration with different groups and world leading companies (IBM, Corning Inc., Oclaro, BASF, SEAT, Zeiss etc.) for tailor-made coatings for different optoelectronic devices including 3D TVs, EUV lithography, nanostructured substrates, head-up display etc.

Master's Thesis Work

Thesis Advisor: Prof. Soma Venugopal Rao (**IIT**, Guwahati) (Now at ACRHEM, Univ. of Hyd, India)

Thesis Title: Nonlinear optical properties of Zn-phthalocyanine and free base phthalocyanine studied using a low power He-Ne laser

Education

- PhD in Photonics, ICFO-The Institute of Photonic Sciences, Barcelona, Spain
- Master of Science (Physics): Indian Institute of Technology (IIT), Guwahati, India
- Bachelor of Science (Physics): Dibrugarh University, Assam, India

Scholastic Achievements/Awards

- **Photonics21 Student Innovation Award**, 2011 by European Technology Platform Photonics21 association. The pan European award was established by Photonics21 Work Group on Research, Education and Training in order to promote research in photonics related to R&D with industrial impact and is awarded annually to a person less than 35 years of age
- **Cum Laude with Honours** (PhD degree)
- **Best Poster Award** at ICFO annual poster session event on Corporate Liaison Day, Barcelona, 2011
- **FPI fellowship** from Spanish Ministry of Science and Innovation (MICINN) to carry out research at ICFO
- Represented ICFO in 5th **India–Spain Dialogue** held at Madrid Spain, 2010
- Qualified Graduate Aptitude Test in Engineering (**GATE**) in 2007, India
- **Merit Scholarship** (IIT Guwahati) for the year 2004–06
- Placed among **Top 10 % Students** in the center of the national level examination in Physics 2002, India

Refereed Publications

- "*Graphene as an anti-permeation and protective layer for indium-free transparent electrodes*," T. L. Chen, D. S. Ghosh, N. Formica, V. Pruneri, **Nanotechnology** 23, 395603 (2012)
- "*High figure-of-merit Ag/Al:ZnO/Ni nano-thick transparent electrodes for indium-free flexible photovoltaics*", D. S. Ghosh, T. L. Chen, N. Formica, J. Hwang, I Bruder, V. Pruneri, **Solar Energy Materials and Solar cells** 107, 338 (2012)
- "Highly stable Ag-Ni based transparent electrodes on PET substrates for flexible organic solar cells," N. Formica, D. S. Ghosh, T. L. Chen, C. Eickhoff, I. Bruder, V. Pruneri, **Solar Energy Materials and Solar cells** 107, 63 (2012)
- "*Efficient polymer solar cell employing an oxidized Ni capped Al:ZnO anode without the need of additional hole-transporting-layer*", T. L. Chen, R. Betancur, D. S. Ghosh, J. Martorell, V. Pruneri, **Applied Physics Letters** 100, 13310 (2012)
- "*Highly stable Al-doped ZnO transparent conductors using an oxidized nickel capping layer at its percolation thickness*", T. L. Chen, D. S. Ghosh, D. Krautz, S. Cheylan, V. Pruneri, **Applied Physics Letters** 99, 93302 (2011)
- "*Organic light emitting diode with indium free metallic bilayer as transparent anode*", S. Cheylan, D. S. Ghosh, D. Krautz, T. L. Chen, V. Pruneri, **Organic Electronics** 12, 818 (2011)
- "*Semi-transparent metal electrode of Cu-Ni as a replacement of ITO in organic photovoltaic cells*", D. S. Ghosh, R. Betancur, T. L. Chen, V. Pruneri, J. Martorell, **Solar Energy Materials and Solar Cells** 95, 1228 (2011) (highlighted in Nature Photonics "News and Views")
- "*Ultrathin Cu-Ti bilayer transparent conductors with enhanced figure-of-merit and stability*", D. S. Ghosh, T. L. Chen, V. Pruneri, **Applied Physics Letters** 96, 91106 (2010).
- "*High figure-of-merit ultrathin metal transparent electrodes incorporating a conductive grid*", D. S. Ghosh, T. L. Chen, V. Pruneri, **Applied Physics Letters** 96, 41109 (2010)
- "*Nickel as an alternative semitransparent anode to indium tin oxide for polymer LED applications*", D. Krautz, S. Cheylan, D. S. Ghosh, V. Pruneri, **Nanotechnology** 25, 275204 (2009) (highlighted in nanotechweb.org)
- "*Stable transparent Ni electrodes*", L. Martinez, D. S. Ghosh, S. Giurgola, P. Vergani, V. Pruneri, **Optical Materials** 31, 1115 (2009)
- "*Widely transparent electrodes based on ultrathin metals*", D. S. Ghosh, L. Martinez, S. Giurgola, P. Vergani, V. Pruneri, **Optics Letters** 34, 325 (2009)

Patent Granted/Applications

- "*Metal Transparent conductors with low sheet resistance*", V. Pruneri, D. S. Ghosh, US2012103669 (A1)
- "*Multilayer metallic electrodes for optoelectronics*", V. Pruneri, D. S. Ghosh, T. L. Chen, US2012260983 (A1)
- "*Transparent electrode based on combination of transparent conductive oxide, metals and oxide*", V. Pruneri, D. S. Ghosh, T. L. Chen, WO2011101338 (A3)

Invited Talks

- *"Highly stable ultrathin Ag-Ni films for flexible transparent electronics",* N. Formica, D. S. Ghosh, T. L. Chen, V. Pruneri, 221st Electrochemical Society Meeting, Seattle, May 2012.
- "*Metallic nano-films for display, lighting and photovoltaic devices*", D. S. Ghosh, N. Formica, T. L. Chen, V. Pruneri, Encontro de Fìsica 2011, Foz do Iguaçu, Brazil, June 2011
- "*Thermally stable ultrathin metal transparent electrodes for solar cell applications*", D. S. Ghosh, T. L. Chen, V. Pruneri, Transparent Conducting Electrodes, Berne, Switzerland, 2010.
- "*Nickel electrodes transparent from UV to IR applications*", D. S. Ghosh, L. Martinez, V. Pruneri, Microtechnologies for the new Millennium 2009, Dresden, Germany, 2009.
- "*Ultrathin metal film: an emerging transparent electrode for the optoelectronics industry*", L. Martinez, D. S. Ghosh, S. Giurgola, P. Vergani, V. Pruneri, 4th International Conference on Advanced Optoelectronics and Lasers (CAOL), Alushta, Ukraine, 2008.

Conference Proceedings/Contributions

- "*Oxidation-free and ultra-smooth thin silver films grown on a copper seed layer*", N. Formica, D. S. Ghosh, A. Carrilero, R. E. Simpson, T. L. Chen, V. Pruneri, CLEO-Europe, Munich, Germany, June 2013.
- "*Avoiding indium in optoelectronics: Oxidized ultrathin metal capped Al:ZnO*", D. S. Ghosh, N. Formica, T. L. Chen, V. Pruneri, Nanophotonics for energy efficiency, Minatec- Grenoble, France, November 2012.
- "*Oxidized Ni capped Al doped ZnO for organic optoelectronics*", D. S. Ghosh, T. L. Chen, N. Formica, V. Pruneri, Nano Transparent Conductive Materials, Minatec-Grenoble, France, June 2012.
- "*Graphene as an anti-permeation and protective layer for indium-free transparent electrodes*", T. L. Chen, D. S. Ghosh, V. Pruneri, E-MRS Fall Meeting, Boston, 2012.

- "*Flexible ultrathin alloy layer as transparent electrode for indium-free organic photovoltaics*", D. S. Ghosh, N. Formica, T. L. Chen, V. Pruneri, International Symbosium on Flexible Organic Electronics, Thessaloniki, Greece, June 2012.
- "*Ultrathin Ag-Ni transparent conductor for flexible optoelectronics*", N. Formica, D. S. Ghosh, T. L. Chen, V. Pruneri, Transparent Conductive Materials, Crete, Greece, June 2012 (Poster).
- "*Al:ZnO capped Ag nano-thick layer for high figure of merit and flexible indium free transparent electrode*", T. L. Chen, D. S. Ghosh, N. Formica, V. Pruneri, International Conference on Frontiers of Advanced Materials and Engineering technology, China, Jan 2012.
- "*Environmentally stable Al-doped ZnO transparent electrode for organic optoelectronic devices* ", D. S. Ghosh, T. L. Chen, D. Krautz, S. Cheylan, V. Pruneri, Renewable Energy and Environment, Texas, USA, Nov 2011.
- "*Fabrication of P3HT-PCBM solar cells using an 8 nm thick semi- transparent metal electrode as an replacement of ITO*", R. Betancur, D. S. Ghosh, X. Elais, M. Maymo, T. L. Chen, V. Pruneri, J. Martorell, E-MRS Spring Meeting, Nice, France, May 2011.
- "*Cu-Ni bilayer Ultrathin Metal Film for conductive transparent electrodes*", N. Formica, D. S. Ghosh, T. L. Chen, V. Pruneri, OPTOEL 2011, Santander, Spain, May 2011.
- "*Ultra-thin metals for indium free organic lighting and photovoltaics*", D. S. Ghosh, T. L. Chen, D. Krautz, S. Cheylan, V. Pruneri, Photonics 2010, Guwahati, India, 2010.
- "*ITO-free organic light-emitting devices using ultra-thin metal films as electrodes*", D. Krautz, D. S. Ghosh, T. L. Chen, V. Pruneri, S. Cheylan, Nanoscale Devices for Environmental and Energy Applications, San Sebastián, Spain, 2010.
- "*Ultrathin metal transparent conductors for energy efficiency*", D.S. Ghosh, T. L. Chen, V. Pruneri, Nanophotonics for Energy Efficiency Plenary Meeting, Grenoble, France, 2010.
- "*Ultra-thin metal electrodes for optoelectronics and photovoltaic*", D. S. Ghosh, L. Martinez, V. Pruneri, EU Concertation meeting in Nanophotonics, Florence, Italy, 2009.
- "*Ultrathin metal transparent electrodes for lighting and photovoltaic applications*", D. S. Ghosh, D. Krautz, S. Cheylan, V. Pruneri, Optics and Photonics for Advanced Energy Technology, Cambridge, USA, 2009.
- "*Transparent Metal electrodes*", D. S. Ghosh, L. Martinez, V. Pruneri, CLEO EUROPE—EQEC 2009, Munich, Germany, 14–19 June 2009 (Poster).

Areas of Interest

- Emerging Transparent electrodes
- Organic photovoltaics and lighting

- Ultra-thin films and their applications
- Graphene and its applications

Personnel Information

- Nationality: Indian
- Date of birth: 28th August, 1982
- Marital Status: Unmarried

Activities and Interests

Apart from research, I enjoy travelling and photography. I do like watching sports especially Soccer, Cricket and Formula 1 and playing indoor games like Chess.

FSC
www.fsc.org
MIX
Papier | Fördert
gute Waldnutzung
FSC® C083411

Kolibri 360 GmbH
Ferdinand-Jühlke-Straße 7
99095 Erfurt, Deutschland
produktsicherheit@kolibri360.de